GARDENS OF HOPE

Gardens of Hope

*Cultivating Food and the Future
in a Post-Disaster City*

Yuki Kato

NEW YORK UNIVERSITY PRESS

New York

NEW YORK UNIVERSITY PRESS
New York
www.nyupress.org

References to Internet websites (URLs) were accurate at the time of writing. Neither the author nor New York University Press is responsible for URLs that may have expired or changed since the manuscript was prepared.

Please contact the Library of Congress for Cataloging-in-Publication data.
ISBN: 9781479827374 (hardback)
ISBN: 9781479827404 (paperback)
ISBN: 9781479827428 (library ebook)
ISBN: 9781479827411 (consumer ebook)

This book is printed on acid-free paper, and its binding materials are chosen for strength and durability. We strive to use environmentally responsible suppliers and materials to the greatest extent possible in publishing our books.

The manufacturer's authorized representative in the EU for product safety is Mare Nostrum Group B.V., Mauritskade 21D, 1091 GC Amsterdam, The Netherlands. Email: gpsr@mare-nostrum.co.uk.

Manufactured in the United States of America

10 9 8 7 6 5 4 3 2 1

Also available as an ebook

CONTENTS

Introduction

Jeanette had been a gardener long before Hurricanes Katrina and Rita devastated New Orleans in the summer of 2005. One of the first things that she did after the storms was to check on her garden. She had prepared the garden for a magazine article just before the Category 5 hurricane made the landfall in the city; now her roses were blooming, as the city faced uncertainty in the immediate wake of the flooding caused by the levee breaches. She recalled:

> I would cut bouquets and take them to first responders, because I felt that so much of the city was dead and gray from the salt water. I thought it would be nice for the volunteers to have a little bouquet to put beside wherever it was that they were sleeping here [in the city]. Because, you know, they were living—a lot of them were living in temporary facilities to help with the cleanup and so forth. So I would go to—they had centers for handing out food, and I would make bouquets and take them there.

For Jeanette, tending to and sharing from the garden was an act of restoring order and demonstrating resilience and hope in times of chaos, disruption, and despair. Over the next decade, she became one of the dozens of urban growers who converted vacant lots into flourishing gardens and farms in New Orleans. They considered gardens and farms to be unique solutions to the problems they saw around them, from blight and food insecurity to environmental destruction and a general sense of hopelessness. But why gardens? And what would it take to start and sustain these growing spaces in the city, especially one that was undergoing rapid major transformations?

Gardening is likely not the first thing that comes to mind for most people confronted with a major natural disaster. In fact, it was not on the minds of most New Orleanians or local public officials in the aftermath of Hurricane Katrina. Nevertheless, a small group of individuals began cultivating vacant lots for growing food across the city, initially in small numbers and gradually expanding in scale and scope. By 2015, more than thirty full-time growers were tending over one hundred new lots across the city. Most of these sites were not traditional community gardens, but instead a new type of cultivation project that took on many forms. They grew vegetables and flowers, raised chickens, sold or donated the food, taught workshops, held events, and watched neighborhood kids play in the garden. But why and how did they do this? This is the story of a group of individuals who saw something inspiring about gardens and farms in a post-disaster city, and an exploration of their triumphs and tribulations as they tried to turn their visions into reality for more than a decade following the 2005 storms, during which some gardens would flourish while others withered. This book follows these growers' trajectories, from their initial aspirations to how they established, maintained, and sometimes terminated their projects. What transpired in post-Katrina New Orleans involved a complex interplay between individual intentions and structural constraints and opportunities, as the growers navigated and adapted to the changing economic and social environments.

There is something about urban gardening that makes us hopeful. In fact, we ascribe so many hopes to the gardens in the city: they will make the city more ecologically sustainable; they will reduce food insecurity in areas that lack access to affordable, healthy food options; they will improve health outcomes by getting people outdoors and encouraging them to eat fresh, in-season produce; they will increase property values by converting blighted properties into well-maintained green spaces; they will create jobs and entrepreneurial opportunities; they will teach children about nature and "where their food comes from." These hopes gain urgency during times of disruption and uncertainty, including natural disasters and public health crises. We expect a garden, which is often no larger than a city lot, to fix so many structural social issues, without pausing to ask: Can it do all that? How would it do all that? Who would make it all happen?

In reality, gardens in the city come and go. Despite their constant popularity, the public rarely considers what it actually takes to start and sustain a new garden in the city, especially when the city is under distress. Historically, urban gardens have emerged during periods of crisis, initiated either by government policy or as a way for marginalized communities to enact "collective efficacy" and cultural preservation.[1] In recent years, a new generation of urban growers have turned to urban agriculture as a form of food justice activism.[2] Yet what I found in post-Katrina New Orleans did not fit these historical patterns of urban cultivation in terms of who was growing, how they were growing, and why they were growing in the city. The individuals who began new gardens and farms since 2005 came to the practice with a variety of aspirations and experiences, and they engaged in urban cultivation not as their leisurely pastime, but as their full-time commitment. They did not perceive their gardens and farms as a panacea, but nevertheless approached their practice as a way to enact their vision of an alternative future, even if their practices did not fit into existing legal, social, and economic systems. In manifesting these visions, their focus remained on immediate, direct action in their cultivation space. Most of the growers seemed uninterested in organizing themselves into a collective movement, even after years of working alongside each other, and despite the robust social justice movements that had always been active in the city. This book unpacks the intentions and actions of these new urban growers by situating them in the shifting economic and social contexts of the post-disaster city. I theorize that these growers were engaging in what I call "prefigurative urbanism," a distinct form of civic engagement, and I explore what they saw in urban cultivation as the site for potential change, how their practices evolved as the city transitioned from recovery to redevelopment, and why cultivation projects took such divergent trajectories, rather than cohere to a larger, uniform movement.

This book focuses on a specific way of growing in the city that I refer to as "urban cultivation." Urban cultivation is not defined by what it grows. It includes planting and harvesting of vegetables, herbs, fruits, or flowers, but it also includes tending of animals, insects, and soil to produce food products and value-added goods, such as pickles, spice mixes, or soaps, or to enhance ecological sustainability. Nor is it defined by a garden's size or form. Urban cultivation can happen in a backyard or on a whole city lot. But not all gardens and farms in the city would qualify as urban cultivation. Most importantly, the urban cultivators in this book each practice engagement with other community members by intentionally sharing food, space, labor, or ideas. Urban cultivation projects must defy or expand the normative expectations of an urban lifestyle and a capitalistic valorization of land, labor, and food. For example, under this definition, a community garden that operates mostly as a private club of individualized plots is not a form of urban cultivation, unless the garden practices some alternative land use or food distribution system. Similarly, an urban farm would not be considered a form of urban cultivation if it operates solely as a business that hires employees to grow, market, and sell its produce. But a private-school garden that invites community members to be stewards of the space alongside students and teachers would be considered a form of urban cultivation, because it blurs the boundaries between public and private space and expands the definition of education to beyond the classroom. In short, urban cultivation must entail more than just the growth of plants, animals, or soil—it must include the cultivation of some new ideas, praxes, or connections.

As I've conceived it, the concept of urban cultivation redirects focus away from the commercial or private production of goods implied in the concepts and practices of "farming" or "agriculture" toward social changes and relationships made possible through the act of tending land and nature collectively. My development and use of this term result from my conversations with the growers for this book, who grappled with the dichotomy of garden and farm terminology, because they felt that neither accurately described what they did. And as a researcher, I found the definition of "urban agriculture" to be both too broad and too inconsistently applied within existing scholarship. Most social scientific studies of "urban agriculture" have either focused on traditional models of community gardening or for-profit farming in the city. For urban cultivators, growing food in the city at scale, especially as one's primary occupation, requires a commitment of time and resources beyond those needed for typical backyard gardens, community gardens, or school gardens. By contrast, natural science research on "urban agriculture" has exclusively focused on biological or ecological aspects of the practice without considering how social dimensions of urban life, such as gentrification, may shape urban growing practices. By introducing the term "urban cultivation," I hope to bring our attention to the social implications of growing food, plants, and soil in the city, so that we can begin to distinguish different forms of gardening and farming for a more critical assessment of their distinctive roles in city life.

Gardening in Times of Crisis

Urban gardening is seeing a revival across the United States. The number of community gardens has soared in most American cities over the last couple of decades, with many having years-long wait lists for a plot. City schools are incorporating gardening into their formal and informal curricula to teach students about plants, nature, and nutrition. But another form of growing food in the city is also gaining prominence: growing food at scale. These practices, often called "urban farming" or "urban agriculture," entail growing food not for personal consumption or for education but for the consumption of others. Food may be donated or sold to individual consumers, restaurants, or grocery stores. Images of young farmers tending rows of crops amid the bustle of the city, rooftop farms overlooking skyscrapers, or rows of shelves lining climate-controlled repurposed shipping containers embody utopian visions of better food systems, a better environment, and better health for all. This is not merely a niche trend. With the passage of the 2018 US Farm Bill, the US Department of Agriculture established the Office of Urban Agriculture and Innovative Production, recognizing and legitimizing the practice of scaled-up food production in cities.

Yet this is hardly a new phenomenon. In most American cities, people have always grown food, not just for themselves, but also for others.[3] These cultivation sites were often located in marginalized communities, tended by growers who were people of color, immigrants, and working-class folks, and remained unnoticed by those who lived outside of these communities. Immigrant growers planted produce, as well as traditional and medicinal herbs, from their native lands that they could not find at American supermarkets, and they often gathered in these spaces to socialize and organize. For immigrant communities, urban cultivation was a way to transplant their cultural heritage by recreating familiar food and social spaces.[4] At the same time, the structural impact of racial segregation, economic disinvestment, and political disenfranchisement led to the abundance of vacant lots and acute food insecurity in low-income communities of color.[5] Thus, for communities of color in post-industrial cities like Detroit and Cleveland, since the 1970s, growing food has always been more than a means to feed themselves; it has also been a form of collective advocacy for reclaiming their right to the land.[6]

Hyperlocal production of food in the city did not occur exclusively in marginalized communities. During the 1960s, some white, middle-class urbanites became supporters of the alternative food movement. Their concerns about the environmental harms of conventional agriculture led them to seek smaller, local food systems, including community gardening, as a small-scale demonstration of their values and an opportunity to raise public awareness.[7] These alternative food movement advocates also saw their practices as politically significant, but their main concerns were notably distinct from that of their immigrant, working-class counterparts in other areas of the cities. Despite the middle-class growers' relative positions of privilege, these cultivation practices remained on the margin of mainstream cultural and social norms, because they existed in decidedly countercultural spaces at a time when "environmentalism" was considered a radical position.

One thing that distinguishes the most recent surge of popularity in growing food at scale in the city is that it involves younger people, many of whom are college-educated, and rarely with direct familial roots in agriculture. These new urban growers come into the practice with a variety of interests and experiences. Many are committed to issues of environmental sustainability and climate change. Others are politically engaged and entered urban agriculture by way of social justice activism,[8] like the increasing number of Black and indigenous growers who approach their practice as a way of reconnecting with the land and their ancestral roots.[9] There has also been an increasing presence of entrepreneurial urban growers whose interests are less politically defined. But why and how did the practice gain new interest in the ways that it has, and are these practices here to stay? To answer these questions, we must pause and consider *when* and *why* certain types of gardens and farms thrive or struggle in a city, *who* are drawn to the practice, and *what* role these gardens and farms play in their communities, beyond providing locally grown food and ecological benefits.

Theories of political economy offer one explanation for the cyclical popularity of cultivation in American cities since the late nineteenth century. During the bust cycle of a city's development, properties lose significant market value.[10] To limit the further decline in property value and to boost public morale, national and local governments have implemented publicly funded, largescale urban gardening programs across US cities, from Detroit mayor Hazen Pingree's potato patches in the 1890s to the urban war garden projects during World War I and II. These crisis-response gardening programs were intended to be temporary, cost-effective approaches to urban land management and, at the same time, serve as patriotic propaganda. In the years following both world wars, many of these garden programs lost funding, gardeners lacked time to tend them, the land became fallow, or the gardens were paved over for redevelopment. These developments continued despite community-based activists' mobilization to save their community gardens and urban farms.[11] For much of the twentieth century, urban planners and political leaders saw gardens and farms in the city as the anthesis of development, and as such, urban cultivation was only recognized as valuable when the land itself held little economic value.[12]

Urban gardening and farming have increased in popularity across the United States over the last couple of decades in two distinct urban contexts that diverge from historical precedent: post-industrial redevelopment and gentrification. In post-industrial, "shrinking" cities, where low-income residents had long grown food for themselves, urban development professionals recognized urban gardens as a way to reactivate vacant lots for development while simultaneously attracting the younger, educated newcomers who were increasingly invested in environmental sustainability and alternative food movements.[13] In cities that continued to gain population during this same period, renewed public and private investment in urban greening, including urban gardening and farming, occurred alongside an in-migration of capital and a whiter, wealthier population that displaced long-term residents and businesses, a phenomenon known as gentrification.[14] Urban planning and policies began embracing "sustainable development" projects, even as the definition and implementation of the term varied widely in practice.[15] In other words, city planners and policymakers in both shrinking and growing cities found new value and potential avenues for redevelopment in urban gardens and other green spaces.

In response to these trends, academics and activists have coined such terms as *ecological gentrification*, *green gentrification*, and *environmental gentrification* to problematize what they describe as the co-optation of greening as a tool to justify development, especially in communities of color or low-income communities that have been identified as gentrifiable.[16] By using the language of sustainability, critics argue, developers and city officials have *greenwashed* the traditional "growth machine model" of urban development that aims to increase property values.[17] Removing an unhoused population to create a sustainable greenway, for example, puts progressive urbanites at ease while also contributing to speculative investment around the proposed greenway.[18] That said, cultivation projects in the city are rarely incorporated into these types of sustainable development policies,[19] suggesting that the practice's increasing popularity may not be the direct outcome of top-down efforts to co-opt gardens and farms for further development. Thus, while theories of political economy help us understand when and why land becomes available and even promoted for cultivation, it fails to explain who would actually end up growing in the city, especially without government sponsorship and public funding.

The most recent wave of urban agricultural activities in the US partially reflects grassroots activists' embrace of gardening and farming as a part of their mobilization tools.[20] Younger urban farmers with some college education who focus on social justice or environmental sustainability in their practice have received scholarly and media attention. Over the last decade, food justice scholarship has examined the potential of urban gardens as sites of *prefigurative politics*, or a form of action-oriented activism that aims to manifest a future beyond what the current social system affords.[21] These studies demonstrate the unique ways in which urban gardens and farms have become places for political organizing, and sites for activists to implement alternative ways of food procurement, land use, and communal ownership. In recent years, media portrayal has gravitated toward these political gardening projects or entrepreneurial projects that market a utopian future of localized food production made possible by technological advancement. Yet these images and stories leave out cultivation projects that are not intentionally and explicitly political, or for-profit farms that seek alternatives to capitalism. Engaging in agricultural practice in the city is in some ways political by nature, because such land-use challenges the prevailing capitalistic principles of the "growth machine."[22] Nevertheless, in the case of post-Katrina New Orleans, to the extent that urban growers exhibited a politics, it was a heterogenous, individualistic, ends-effacing politics, as they demonstrated alternative land uses, helped created locally-grown food distribution systems, and established "growing" as an urban occupation. This book aims to uncover what motivated these individuals to start gardens and farms in a post-disaster city and to document their journeys over time.

Urban Cultivation in Post-Katrina New Orleans

New Orleans has long been portrayed through contradictory narratives of being simultaneously generalizable and exceptional.[23] In the months and years following the levee failure, these opposing narratives would be amplified by both the national media and scholarly research. Numerous books, journal articles, newspaper and magazine articles, blogs, and documentary films have recorded and analyzed *what went wrong* in the city, from the failed federal and local responses to disaster profiteering during the recovery. These works tended to situate the experiences of post-Katrina New Orleans and its residents as exemplars of the impacts of capitalistic greed, racial inequity, and the ever-shrinking welfare state on American cities.[24] Critical analyses of the disaster recovery process focused especially on how the events surfaced long-term patterns of systemic racism that put the city's low-income African Americans at a significant disadvantage in their ability to evacuate, survive, and return.[25] Counternarratives chronicled the grassroots activism and the role of community-based institutions to resist the neoliberal forces of redevelopment and to preserve the *soul* of the city.[26] These narratives emphasized the resilience of New Orleanians and the city's unique cultural traditions to demonstrate the place's exceptional status, an American city like no other.

The polarized depictions of post-disaster New Orleans reflected varied expectations for the future of the city. Politicians and businesses saw an opportunity to revive the city's long-stagnant economic infrastructure by taking advantage of the recovery funds and the new level of national and international attention paid to the city.[27] Residents in predominantly Black and working-class neighborhoods organized themselves to fight to preserve their community, culture, and memory, often by highlighting the essentiality of their contribution to the city's economic and cultural assets. The influx of mostly white, college educated newcomers, whose rate of arrival accelerated after 2012 with many seeking to be a part of the new economic development, enhanced the sense of urgency about the loss of an authentic or exceptional New Orleans,[28] just as these concepts became central to the city's business and tourism marketing.[29] Local food, music, vernacular, traditions, and flare were claimed by both the economic redevelopment forces and the community activists as their symbolic raison d'être—though of course each side held its own definition of what *locality* meant and to whom it belonged. In this context, urban cultivation aligned itself with neither camp. Growers occupied an ambiguous space between economic growth and environmental sustainability, and between neoliberal governance and grassroots activism, as they adapted to the shifting political and economic environments of the post-disaster city.

As the water receded and the long-term rebuilding process began, it became clear that the *new* New Orleans was going to be shaped by neoliberal policies that emphasized free-market solutions and individual responsibility, along with narratives of experimentation and innovation that positioned the city as a "blank slate" waiting to be reinvented. Sociologists Kevin Gotham and Miriam Greenberg studied the post-disaster transformation of post-Katrina New Orleans and post-9/11 New York City and found that urban crises play a catalytic role in cycles of legitimating and intensifying neoliberal urbanism, whereby market-driven responses to the crisis exacerbate, rather than alleviate, vulnerabilities through uneven investment in social and physical infrastructure.[30] Urban cultivation was not a part of this recovery agenda. The practice remained overlooked for the first half of the decade by the city and state governments, the real estate market, and businesses. It was only after the urban cultivation scene took root and began to show its strength that government agencies and policymakers took note of the practice and realized that it could be co-opted into the urban redevelopment agenda. The growers' relationship to the city's "neoliberal turn"[31] was complicated, as they selectively took advantage of some opportunities, such as an expanded market for selling their produce, while retaining a level of skepticism toward those attempting to co-opt their practice.

During the early years of the recovery, growers established cultivation projects in hopes of contributing to the community rebuilding process. These projects were initiated primarily by multi-generation New Orleanians or long-term residents of all ages, Black and white, and for the most part, emerged separately from the local grassroots activism advocating for racial justice on a variety of issues from education and housing to criminal justice system. As time went on, the grower profile increasingly reflected the demographic changes in the city, with the cultivation scene dominated by younger, mostly white post-Katrina transplants by 2015, and with the majority of sites operating in predominantly Black neighborhoods that experienced slower rates of recovery. Yet these growers did not fit the typical profile of gentrifiers who avoided engagement with long-term residents. Their daily presence in the community while tending the gardens provided opportunities to meet and interact with the nearby residents on a regular basis. Some forged close relationships with neighbors, while others sensed a mixture of curiosity, suspicion, and reluctant acceptance by the long-term residents. At the same time, growers continued to experience difficulties accessing land and resources to start and sustain their cultivation projects, despite their relative socioeconomic privilege, and many projects never materialized or terminated even after many years in operation. The ambiguous space that the growers occupied underscores the need to recognize and understand the types of attempted social change that are often overlooked by social movement scholars and the media.

Urban Cultivation as a Form of Prefigurative Urbanism

Why did these individuals start urban gardens and farms in a post-disaster city if they were not participating in either the formal recovery agenda or grassroots activism? I describe the urban cultivation scene that emerged in post-disaster New Orleans as a form of "prefigurative urbanism," or everyday practices that demonstrate alternative ways of life that do not yet exist in the city. I define prefigurative urbanism as a civic action that aims to bring about immediate, tangible changes through direct actions that deviate from social and legal norms. It involves individualized, localized actions that are not organized cohesively, as each actor or group of actors enacts their own visions of the future through a unique set of direct, publicly visible activities. Rapid, rather than gradual, changes are prioritized in prefigurative urbanism. As a noncollective movement, for example, the growers' initial focus is on finding a space to start growing, rather than advocating to legalize urban cultivation or mobilizing around a right to grow in the city. Immediate direct action is the priority, and pragmatism prevails as the guiding principle, as they face expected and unexpected hurdles and shift courses in order to keep growing. The long-term vision is left open, and the actors foresee that their current actions would eventually coalesce into broader changes, but their immediate actions are regarded as being valuable for what they deliver now, rather than as a means for systemic change.

Prefigurative urbanism's emphasis on direct action and prompt social changes may remind some readers of prefigurative politics. Political scientist Carl Boggs defines prefigurative politics as "the embodiment, within the ongoing political practice of a movement, of those forms of social relations, decision-making, culture, and human experience that are the ultimate goal."[32] It is commonly described as the practice of "enacting the world that you wish to see, now" rather than engaging with the procedural politics such as implementing policies or raising awareness in hopes that these changes will eventually bring about desired changes someday. Prefiguration involves imagining and manifesting what could be, or should be, by acting as if the change is *already* here. In the United States, the Black Panther Party's community breakfast program is a prominent example of prefigurative politics. The Panthers' program predated federally funded breakfast programs at public schools, which did not begin until 1966, and demonstrated that children in Black communities should and can go to school with full stomachs and the knowledge that they are cared for. In the agricultural realm, sociologist Monica White's work chronicles historical and current practices of prefigurative politics through farming within Black communities in the US.[33] As White suggests, these cultivation practices redefined and reclaimed agricultural labor and skill as assets available within their communities, and their gardens and farms became political organizing spaces, even if some of these efforts were undermined, suppressed, or short-lived.

Aside from the focus on immediate, tangible actions that bypass procedural politics, there are additional similarities between prefigurative urbanism and prefigurative politics (Table 1.1). Both aim to demonstrate to the public the possibility of change in forms of concrete manifestation of the future in the present, which could be difficult to imagine for individuals skeptical or afraid of change. The enactment of concrete

TABLE I.1. Distinct Characteristics of Prefigurative Politics and Prefigurative Urbanism

Prefigurative Politics	Prefigurative Urbanism
Direct, tangible actions	
Immediate changes over procedural politics	
Public demonstration of alternative social systems	
Experimental and adaptative	
Ends-guided or Ends-effacing	Ends-effacing
Collective	Individual
Explicitly political	Apolitical or implicitly political
Ideological	Pragmatic
Intentionally disruptive	Disregard for disruption

changes, therefore, must take place in the public realm, physically or socially, so that the prefiguration is visible and noticeable to the public. The actors may engage in public outreach through interpersonal or virtual communication, though it is the action, not words of promise, that is paramount. If seeing is believing, then feeding hungry children before school or growing food at scale in a small urban plot convincingly shows that an alternative *is* possible. Finally, these changes must happen in a relatively short span of time for the public to take notice of the transformation.

Despite these similarities, there are a few notable differences between prefigurative urbanism and prefigurative politics. In prefigurative politics, prefiguration may be ends-guided, with a specific and concrete vision of the alternative future to be enacted, or ends-effacing, which prioritizes intentional disruption of the existing system without defined consensus on what the end point looks like.[34] Prefigurative urbanism is driven primarily by an ends-effacing form of prefiguration, which reflects prefigurative urbanism's tendency to be enacted by a heterogeneity of individualized visions and actions. Unlike prefigurative politics, which is a tactic of a collective social movement, actors within prefigurative urbanism typically do not frame their practice as explicitly political, even if it has political implications and outcomes. To this effect, disruption itself is not necessarily the main intention of prefigurative urbanism praxis, but it is worth noting that the actors exhibit a lack of regard for norms and laws. In other words, they are not there to be troublemakers per se, but they are comfortable with bending rules, operating in the legal gray zones, or "asking for forgiveness later rather than asking for permission first" in order to accomplish what they set out to enact. Ultimately, prefigurative urbanism actors' decisions on how to experiment and adapt to shifting social and economic environments are guided by pragmatism over ideology. They are driven by their desire to do "something, now," to bring about *some kind of* change, especially in times and places where there is little trust for the government or other powerful entities to make a real, tangible difference.

Prefigurative urbanism *activates* hope through action, especially in times of uncertainty and despair. Because of its tangible nature and vast capacity to be symbolically meaningful, urban cultivation resonates with those who are drawn to taking concrete actions in order to feel hopeful, even when they are not cognizant of what they are hopeful for or how to implement these changes at a larger scale. To this effect, prefigurative urbanism actors tend to exhibit traits of being social deviants and independents who are not afraid of staying outside the norm and prefer to get things done rather than talking about solutions. Being in some form of liminal, or in-between, period in their lives could also intensify these actors' focus on the present rather than the past or the future.

In fact, disasters and other moments of crisis intensify the popular association between urban cultivation and hope, even without government promotion, dedicated resources, or grassroots activism. For example, "pandemic gardening" became popular during the initial year of the COVID-19 pandemic, especially among the middle-class urbanites who previously were not engaged in growing. The acuteness of food insecurity was palpable as consumers encountered empty supermarket shelves or were forced to endure long food pantry lines. The massive scale of social instability was disorienting to many who were not used to facing such challenges on a regular basis. For those fortunate enough to be able to work remotely, gardening at home was one way to be outside while also taking advantage of the extra time they gained by no longer commuting or socializing with others. But more fundamentally, gardening appealed to these new gardeners at a primal level, offering a way to do something very basic, routinized, and instantly gratifying in the face of a global health crisis. The direct, tangible, and immediate nature of growing one's own food, or gathering with others in the garden, is a convincing manifestation of positive change and contributes to a greater sense of stability. To be sure, this example from the early months of the global pandemic illustrates that social status plays a significant role in who is drawn to what they perceive to be an alternative future, and what capacity and resources they can access to act on their aspiration.

Gardening can induce hope, but what we aspire to achieve through it is not singular or even self-evident. Gardening can be a way to regain and demonstrate one's sense of self-efficacy, especially when the solutions to problems in other aspects of one's life seem uncertain or beyond one's control. For those of us whose familial lineage or collective memories are closely connected to agriculture or local food provisioning, experiencing the joys and struggles of farming is a way of honoring history and connecting to our ancestors and the land.[35] Being in the garden, a form of nature, can be a refuge from congestion and urban architecture, particularly for people without access to private open spaces like backyards. It provides a temporary but real space where growers can exercise a sense of freedom and control, even if it is not articulated explicitly as a political gesture.[36] Working with one's hands and getting *dirty* may also appeal to younger urbanites who increasingly seek *authentic* or *alternative* careers and lifestyles.[37] While all of these forms of hope have political and economic implications, the kinds of change that urban growers in post-Katrina New Orleans were trying to enact were not necessarily political or even fully articulated by the growers themselves. Yet they grew not because they were hopeful, but rather, in order to cultivate hope through their actions.

Hope is not simply a state of mind, but it is deeply connected to one's sense of agency. Philosopher Victoria McGeer distinguishes "willful hope," the way we stake our optimism on our own capacity to bring about needed change, from "wishful hope," which relies on external forces to deliver these changes.[38] McGeer cautions against leaning too strongly into either type of hope, as doing so may result in disappointment and despair when individual actions are not sufficient to bring about desired changes, or when the help from outside never arrives.

Instead, she suggests we should strive for a "responsive hope" that balances the two types of hopes through interaction with others. She writes:

> In hoping, we create a kind of imaginative scaffolding that calls for the creative exercise of our capacities and, so often, for their development. To hope well is thus to do more than focus on hoped-for ends; it is critical to take a reflective and developmental stance toward our own capacities as agents—hence, it is to experience ourselves as agents of potential as well as agents in fact.[39]

In the case of urban cultivation projects that developed in post-Katrina New Orleans, starting new gardens and farms was an act of willful hope, one in which growers tangibly manifested their visions for positive change. Doing something *now* defies both pessimism and despair in the moment. Over time, however, growers' willful hope is challenged by the hurdles they face in trying to bring into existence something that does not fit squarely into existing social and legal norms. In these instances, growers' pragmatism and their focus on direct action allowed them to persevere. In other words, it was not their optimism that kept them going; their commitment to continue their work maintained their hope. As philosopher Terry Eagleton notes in his work dissociating hope from optimism, "the mere act of being able to imagine an alternative future may distance and relativise the present, loosening its grip upon us to the point where the future in question becomes more feasible."[40] The urban growers that this book focuses on were continuing to imagine and enact alternative futures through their concrete action, thus sustaining their hope, precisely because they had little trust in the existing system to bring about meaningful changes.

By situating the experiences of individuals in the context of social structure, this book explores why urban cultivation in post-Katrina New Orleans emerged as a form of prefigurative urbanism and how the practice changed over time. Both the city and the individuals in this study experienced rapid and drastic changes over the course of nearly two decades between 2005 and 2023, and growers' practices adapted to these changing circumstances. We will also see changes in who was growing at scale in the city, and contemplate why some cultivation spaces continued to expand and evolve while others faltered. This book focuses on the relationship between external political and economic conditions and individual actions to understand what motivates people to do something that they view as transformative, and why some of them choose to enact their visions through a distinctly independent, tangible, and exploratory mode of civic action.

The Study

I trace the beginning of this project to my early years of living in New Orleans, around 2008. As an urban sociologist, I had a broad curiosity about the gardens and farms that were slowly appearing across the city. I wanted to know why people were starting gardens and farms, so I began talking to these individuals, who I later learned called themselves "growers." Starting in 2010, I conducted several preliminary research projects to make sense of the city's emerging alternative food movement scene, before launching a formal research project in 2013 that focused on land access and retention among urban cultivation project operators. Graduate students Cate Irvin and Scarlett Andrews helped me interview fifty individuals between 2014 and 2015, the majority of whom were still engaging in what we were describing at the time as "urban agriculture" in New Orleans. The study also included interviews with seven individuals who had already left the practice or the city by then. In addition to interviews with growers, I interviewed ten leaders of local nonprofit organizations and representatives of public agencies whose work directly impacted urban cultivation.

I moved from New Orleans to Washington, DC, in 2015, which concluded my initial data collection period. But I kept in touch with a few of the growers; I corresponded with them from time to time to get some sense of the urban cultivation scene in the city, as intense gentrification continued in the city. In 2018, I learned that the alternative food market that I once studied had closed and that a few of the growers that I had interviewed had left the city. I decided then to reach out to other original interviewees to conduct follow-up interviews, and I learned that many of them had also left their project or the city for various reasons, though some were still going strong. Altogether, between 2018 and 2020 I conducted follow-up interviews with twenty-four of the original fifty participants and one new interview.[41] Taking my time with the study has had its merits, because I was able to witness the changes over the decades in real time. In 2023, I had a final check-in conversation with eleven of the follow-up interviewees, all but one of whom was still growing in New Orleans at the time.

The growers that agreed to be interviewed for this study come from a range of demographic backgrounds and brought to the practice of urban cultivation a wide set of experiences. During the initial interview recruitment process, our aim was to locate as many different types of urban cultivation projects as we could, including failed attempts, to gain insights into the range of opportunities and challenges growers faced as the city transformed before our eyes. Of the 51 growers that we interviewed, 25 identified as female and 2 as non-binary. 37 identified as white,[42] and 11 identified as Black or African American. 14 of them described themselves as having grown up in New Orleans and its suburbs or having direct family roots in the city. The same number were not "from New Orleans" but had already been living in New Orleans, some for more than twenty years and others only for a year, before Katrina struck in 2005. 22 of the original grower interviewees came to New Orleans after 2005, some as a part of the initial recovery efforts and others later on, for reasons unrelated to the disaster or urban cultivation.

Many of the individuals named in the book have graciously agreed to be recognized by their actual first names while others remain anonymous.[43] All cultivation projects are referred to by their real names. Throughout the book, references to specific individuals or projects are meant to illustrate particular theoretical concepts or common experiences and are not intended as a holistic depiction of a grower's work or of the project. The reality of growing food in the city is complicated, and the growers and cultivation projects were multi-dimensional and everchanging. All quotes and descriptions should be understood as snapshots of a particular place and time, and only as an aspect of their experiences, observations, or opinions of a particular time and occasion.

To gain a deeper understanding of the city's history of urban cultivation, I also conducted archival research at Tulane University, the New Orleans Public Library, the Historic New Orleans Collection, and the Notarial Archives of Orleans Parish Civil Clerk of Court, with a focus on who used to grow food in the city, why, how, and for whom. The search led me to a range of relevant issues from the region's agricultural economy, its complex and persistent racial politics, and the economic cycle over its 300-year history.

This project was an inductive research journey. For over a decade, I continued to revise my research questions as I observed the remarkable changes taking place in New Orleans as a whole and within the urban cultivation scene, supplemented through my reading of the scholarship on environmental justice and food justice. I am not a grower or an activist, and I considered myself an academic expert on neither the environment nor food when I started the project. Moreover, as a Japanese immigrant who had relocated to New Orleans to take an academic position in 2008, I had very limited knowledge of the city's history or culture. I was an outsider to what I was studying in so many ways, and I was fully aware of this throughout the process. Sometimes, I used my outsider status to my advantage, in that it allowed me to ask naïve questions that someone with insider knowledge might not have posed. My outsider status also allowed me to approach the study with relatively few presumptions or convictions about what urban cultivation is or should be. But being an outsider certainly had its constraints. Some individuals may not have been as candid or transparent in their interview responses as they would have been if I were "one of them." It is also possible that I failed to ask about or notice some aspects of their experiences because I was not actively engaging in urban cultivation or prefigurative urbanism myself. I tried to corroborate interviewee statements with external sources whenever possible, and I hope that conducting follow-up interviews has reduced these negative impacts of my outsider researcher status.

This book does not categorically promote urban cultivation, nor does it merely offer yet another critique of neoliberal urban development. My aim with this book is to serve as an outside observer of the scene, because that is who I am, but also because I believe the perspectives of those operating in more ambiguous political and economic spaces have been lacking in contemporary accounts of urban cultivation in the media and the scholarship. The growers I studied were not perfect individuals. Some started their projects without sufficient preparation and faced hurdles and failures as a result. Most made mistakes along the way and experienced steep learning curves, as did I in my own journey as a researcher on this topic and as a post-Katrina transplant. But they believed that what they were doing was meaningful, and they showed up in the heat and humidity of Southeastern Louisiana each day to walk the walk, quite literally. I have enormous respect for what they tried to accomplish, though as a researcher I feel it is my responsibility to analyze and present all aspects of what happened with the urban cultivation scene in post-diaster New Orleans, so that all of us gain a more grounded and nuanced understanding of when and how urban cultivation and prefigurative urbanism *work* in the city.

Plan of the Book

To contextualize the development of urban cultivation since 2005, the first chapter takes readers back in history to explore the cyclical popularity of urban cultivation in New Orleans. While some people in the city have always grown food, the practice had become less commonplace by the time Hurricanes Katrina and Rita passed through the region in 2005. The remainder of the chapter highlights key events following the storms that are essential for readers' understanding of the trajectory of the urban cultivation scene. I divide the post-Katrina decade into three periods: recovery (2005–2008), transitional (2009–2011), and redevelopment (2012–2015). The chapter underscores how the development of urban cultivation took place in the city's rapid and drastic social transformation over these periods. Changes in who started new urban cultivation projects in the city and how, what, and for whom they were growing reflected these citywide changes. This chronological overview becomes a core thread that runs through subsequent chapters, wherein I elaborate on what motivated the growers to start urban cultivation practices in the first place, and how their attempts at starting and sustaining their projects unfolded over time.

Chapter Two explores growers' visions for the alternative future of the post-disaster city by developing a typology of urban cultivation aspirations. The chapter demonstrates the heterogeneity and multiplicity of aspirations among the growers, as any given grower often held multiple aspirations and added or shifted their priorities over time, both in response to their personal interests and capacities as well as to the transformations happening in the city. I argue that urban cultivation served as a *cultural repertoire* for growers to manifest their visions for alternative futures and to enact willful hope. Regardless of what initially inspired the growers to pursue urban cultivation, putting their aspiration into practice was the first test of their willful hope.

Chapter Three presents the range of challenges and opportunities that growers faced in their attempts to start new urban cultivation projects in the post-disaster city, including land access, permitting, and horticultural skills. The chapter traces the changing nature of these challenges over the decade. When faced with these hurdles, the growers primarily opted to find solutions for their immediate problems rather than form a movement around their shared grievances. Growers distanced themselves from efforts to co-opt their practice into the capitalistic redevelopment agenda while also taking advantage of new opportunities that came with the neoliberal remaking of the city.

How do gardens and farms enact prefigurative urbanism? Chapter Four illustrates the diverse range of practices that the growers engaged in through their cultivation projects to highlight the distinct characteristics of prefigurative urbanism: as they show, it is exploratory, experimental, and adaptive. Its ends-effacing nature allows the actors to stay focused on direct, immediate action, adjusting their practices as they go while continuing to seek alternative forms and meaning by doing rather than theorizing or planning. The prefigurative urbanism approach resulted in wide-ranging trajectories across cultivation projects as the growers independently pursued their own paths.

Over the decades, some gardens went fallow for just as many reasons as they got started. Chapter Five offers a cautionary tale of urban cultivation as a form of prefigurative urban practice by analyzing when and why some projects never took off, struggled, or terminated even after years of operation. The transformative impact of urban cultivation is most observable in a relatively short time span—within the first few months and years—but sustaining gardens and farms long-term requires a different set of skills, resources, and motivations. Prefigurative urbanism's emphasis on direct action without a larger "end point" in mind poses unique set of constraints for the practice's longevity. The chapter also considers the racial and class privileges that affect who can enact prefigurative urbanism and whose practice gains public recognition for its prefigurative potential.

Chapter Six offers a theoretical consideration of the distinct characteristics of prefigurative urbanism and its potential to effect changes to the way we live, work, and experience spaces and places in the city. The chapter situates the concept in the current scholarship in urban studies, disaster recovery, social movements, and environmental and food justice; it presents prefigurative urbanism as a departure from the prevailing narrative in these studies that focus on two opposing forces: neoliberal power elites and grassroots community activism. It contemplates how we should evaluate the significance of prefigurative urbanism in the city, and why this particular form of civic action resonated with a select group of individuals in the post-disaster city. This final chapter also discusses policies nominally designed to promote urban agriculture, or prefigurative urbanism more broadly, and reflects on the implications of this book for aspiring urban growers, those currently engaging in urban cultivation, and those who want to support it.

The mixed outcomes of the urban cultivation scene in post-Katrina New Orleans compels us to question why and how we should sustain gardens and farms in the city. If urban cultivation should be a permanent fixture in the city, then who should sustain it, and how and for whom should it be retained? The implications of this book extend beyond New Orleans and apply to cities that are undergoing major transformations, from gentrification and economic restructuring to climate change. Across this book, we'll explore what Pamela Broom, a lifelong urban cultivation advocate in New Orleans, often describes as "the good, the bad, and the ugly of urban agriculture."

1

After the Rain

The Rise of Urban Cultivation in Post-Katrina New Orleans

On June 13, 2015, NOLA Social Ride and Parkway Partners hosted the fifth Urban Farm Bike Ride. The bicycle social club, founded in 2010, had established the urban farm tour as an annual free event in association with the Eat Local Challenge, a month-long commitment to only eat locally grown or processed foods during the month of June. On the tour, dozens of participants rode their bikes across town, making stops at three urban farms and a community garden in different neighborhoods. The article on the bike tour event published on nola.com, the online platform of the *Times-Picayune*, began as follows:

> Tons of steel and miles of concrete have gone into the rebuilding of New Orleans in the decade since Hurricane Katrina. Lot of seeds and soil also have been a part of that rebirth. A growing network of urban farms is flourishing across the city with green spaces sprouting produce, chicken coops and herb gardens wedged between residential lots, next to schools and on formerly blighted properties.[1]

The idea of a "rebirth" of the city connoted an establishment of new life and norms, rather than a resurrection of the past. The article underscored the novelty of urban farming in the city with a quote from the executive director of a local nonprofit organization, claiming that there was only one urban farm prior to Katrina but "now, there are thirteen." The photos accompanying the article showed the tour participants in a leisurely ride around town, walking through the farms, and standing in a circle listening to the grower, alongside some close-up shots of peppers, berries, and flowers.

The tone of the article's text and images exemplified the positive public reception of urban farms in New Orleans at the time. The narrative of the farms flourishing amidst destruction and abandonment evoked the force of the Mother Nature restoring order and nurturing life. The article made it seem as though urban farms were bound to emerge in the disaster-stricken city's many vacant lots, like the weeds that take over abandoned spaces. But these farms did not just manifest by themselves, and nothing about their beginnings was natural or inevitable. Post-Katrina New Orleans, it turned out, was not necessarily a fertile environment for urban cultivation, especially during the initial recovery period. And nobody could have predicted if it would work or what it would look like as the city underwent extensive transformation. While the proliferation of urban cultivation in the city after the 2005 storms was not orchestrated by any single person or organization in particular, it was not completely accidental, either. So how did a group of people decide to become full-time urban growers at about the same time in the same place? Answering this question requires us to first grasp the social, political, and economic transformations New Orleans has experienced, not only since Katrina but also before the storm, in order to contextualize the kinds of opportunity and constraint that shaped urban cultivation in the city over time.

This chapter will first take us through the history of urban cultivation practices in New Orleans since its early years of French and Spanish colonization through the end of the twentieth century. This history makes it clear that the practice of growing and harvesting food locally was once a common practice among local immigrant and Black populations, though the practice rose and fell in its popularity over time in response to the shifting economic and political conditions of the city. A retrospective historical view is necessary for us to better understand how the development of gardens and farms in New Orleans since 2005 emerged separately from its historical precedents. In *Everything in Its Path*, sociologist Kai T. Erikson traces the history of Buffalo Creek, West Virginia, from its formation to the arrival of a coal mining company, and to the 1972 disaster, when three dams at the mines burst, wiping out the Appalachian coal mining community. Erikson's analysis provides a sociological and anthropological framework for considering how the economic and cultural conditions of the community prior to the disaster shaped the trajectories of the disaster's short- and long-term social impacts.[2] Erikson also makes it clear that the "disaster" involved more than just the flooding that destroyed community infrastructure and killed more than one hundred people; the disaster had begun earlier, and was the result of years of economic exploitation and the destruction of the community's natural resources and social capital, forming the contradictory cultures of self-reliance and dependency that left this community vulnerable. Because of these structures, Erikson argues that the residents' collective and individual trauma cannot be understood outside Buffalo Creek's longer history. Similarly, understanding why urban cultivation emerged as a form of prefigurative urbanism in post-Katrina New Orleans requires us to take a longitudinal approach by focusing on how urban cultivation practices in the city emerged in the past, and why these histories came to be viewed as disconnected from its post-2005 resurgence.

Before the Storm: History and the Memory of Urban Cultivation in New Orleans

New Orleanians of all walks of life used to *eat locally*.[3] People grew and procured food in the city or in the nearby parishes for their consumption. There were vegetable gardens set up at the French Quarter mansions. Working-class folks learned to fish for crabs in the bayou from their parents, or had kitchen gardens to grow tomatoes and herbs. Orchards throughout the city offered a bounty of kumquats, lemons, and pecans every fall and winter. To this day, seasons and holidays are associated with local ingredients and dishes, from king cakes for Mardi Gras, to crawfish boil in the spring, Creole tomatoes in the summer, and satsuma citrus in the winter. The city's diverse population and cultural heritages produced unique culinary traditions that combined the techniques, flavors, and ingredients of Indigenous peoples with influences from Europe, the Caribbean, Africa, Latin America, and more recently, Vietnam. Dishes such as shrimp Creole or gumbo have been marketed as unique and *authentic* culinary offerings in the tourist-centered fine dining establishments,[4] but are also consumed by residents as a part of their folk foodways. Many restaurants and cafeterias across town continue to serve red beans and rice as the daily special every Monday. But most of the ingredients for these beloved local dishes are no longer grown locally, including what is known as the holy trinity of Cajun and Creole cooking: celery, onions, and green peppers.[5]

There is ample evidence of urban cultivation going as far back as the early days of colonization of what eventually became New Orleans, first by France and then Spain. Landscape architect Lake Douglas describes in detail the types of gardens and orchards that were kept in private homes in New Orleans in the eighteenth and nineteenth centuries, remarking that:

> Growing plants for the table and for medicinal purposes was far more important than growing plants for decorative and ornamental purposes, though, of course, many will argue that there is beauty in a well-kept kitchen garden. In New Orleans, as elsewhere in colonial America, it was only after a certain level of economic, political, and social stability was reached that gardeners turned their garden spaces and horticultural attention from function to ornament.[6]

Douglas's inclusion of images from the Plan Book of Plans, from the New Orleans Notarial Archives, visually confirms that many building plans for homes built in the nineteenth century included vegetable gardens and orchards on their property, in addition to ornamental gardens. In the introduction to her translation of Jacques-Felix Lelièvre's 1838 *New Louisiana Gardener (Nouveau Jardinier)*, Sally Reeves points out that the book's audience was likely urban or suburban gardeners, rather than large-scale plantation growers, based on its emphasis on techniques for planting and harvesting vegetables, herbs, and fruit trees.[7] In Louisiana, growing and selling produce gave enslaved people limited yet concrete economic opportunities. A part of the 1795 Decree issued by Governor Carondelet during the Spanish rule "gave official blessing to the system of slave self-provisionment by affirmatively urging masters to assign fields to their slaves for their own cultivation and use,"[8] which allowed enslaved people to grow and sell produce and crafts on Sundays in Congo Square, just outside the French Quarter, on their day off from plantation labor. Earning income through such trade gave these individuals a financial opportunity to buy their freedom out of slavery.

After the Civil War and the abolition of slavery, in response to shifting labor conditions and agricultural scale, larger rural plantations were subdivided. Smaller farms began to operate on these subdivided lots, initially in rural land and eventually in Orleans Parish's suburban subdivisions. They came to be called "truck farms," and farmers sold their produce at local markets, in addition to distributing their products nationally, thanks to the invention of refrigeration cars.[9] Italian immigrants' engagement with truck farming has been better documented,[10] but Black truck farmers were a part of this form of urban commercial cultivation since its origins.[11] The latter half of the 1800s was a time of population and economic growth for the city, with European immigrants arriving to work in the port and the rural Black population leaving or being expelled from the exploitative sharecropping in the country.[12]

New Orleanians across the city grew food at home on their front porches and in their backyards during the first two centuries of the city's history. In an interview archived by the HistoryMakers, Paul R. Valteau Jr., a white former Civil Sheriff of Orleans Parish, recounted New Orleans during the Great Depression as described to him from his mother:

> As you know, New Orleans, during that period of time, was a, a city of mixed neighborhoods and mixed neighbors, and one of the famous doctors over the years here in town grew up in the yard next to their yard—the two houses abutted, and he often talked about how they swapped food that—my grandfather had a garden next to his house—a large garden next to his house where he grew vegetables. And he would share the vegetables with the neighborhood in exchange for meat that they would get to—so that they would have balanced diets.[13]

As his secondhand memory describes, subsistence farming was not exclusive to one social group, and was often practiced communally. Horticultural knowledge was passed on generationally from their trades prior to arriving to the city, both among the European immigrant and Black communities. The folk foodway of procuring food locally thus served the economic purpose of extending the food budget and the cultural purpose of honoring one's ancestral heritage. The practice continued through the first half of the twentieth century, as more of the Black population migrated out of the rural South in what would eventually be known as the Great Migration,[14] with some of them settling in New Orleans.[15] But by the years following World War II, the communal sharing of homegrown food in the city's predominantly Black

neighborhoods was no longer a common sight for Mr. Valteau's generation, even though the need for food increased, and the availability of
land decreased.

During the two world wars, the federal government instituted urban
gardening programs to boost food production and morale back home,
especially among women who were defending the home front. The National War Garden Commission, originally established in 1917 as the
National Emergency Food Garden Commission, encouraged the implementation of school gardens, as well as the conversion of vacant lots
or backyards into gardens by national and local women's clubs.[16] War
gardens, which came to be called "victory gardens" as the war neared
its end, were active in New Orleans, where "approximately 290 acres of
land, comprising 11,000 individual gardens of average size 25 by 25 feet,
were cultivated during the year 1917."[17] After the First World War ended,
the enthusiasm for gardening continued, but with the federal funding
redirected elsewhere, only a small number of gardens were maintained.
A few decades later, the prospect of the Second World War reignited
public interest in small-scale urban gardening. As a result, the federal
government promoted victory gardens again through state extension
agencies, while local women's clubs and the National Victory Garden
Institute each played a significant role in this national effort.[18] But once
again, the funding dissipated after the war, and many of these gardens
across the country struggled to continue despite the desire of many gardeners to carry on. New Orleans was no exception in this trend.

During the second half of the twentieth century, urban cultivation
practice in the city reflected significant transformation in its economic
and social conditions. Hurricane Betsy's major devastation in 1965
marked the city's first population decrease in the century. The racial integration of education and commerce in the 1960s in response to the
Civil Rights Movement accelerated outmigration of white populations
and white-owned businesses into suburban parishes, leaving a much
more racially and economically segregated city behind. For example,
historian Juliette Landphair reports that "from 1940 to 1970, the non-
white population of the area rose from 31 percent to 73 percent, and by
1970, 28 percent of Lower Ninth Ward families lived below the poverty
line" as the crime rate soared in the neighborhood.[19] The oil industry's
collapse and the crack epidemic in the 1980s further accelerated the

white and middle-class exodus from the city, causing a concentration of blight and poverty in predominantly Black communities.[20] As properties were abandoned and food insecurity intensified, this period saw optimal conditions for a resurgence in urban cultivation. But backyard and communal gardening became increasingly less prominent during that time, with the exception of private kitchen gardens.

Declines in truck farming resulted from various coinciding factors, from the modern supermarket system's rising dominance, to racial integration of commerce, the loss of land to urban development, and the bureaucratic formalization of the truck farming business.[21] This period also saw the introduction of faster, cheaper, and more processed food, such as packaged goods and fast-food restaurants, that changed food provisioning practices in the city's low-income communities. The older generation who had knowledge of and experience in growing food at scale were aging out of the practice, and they encouraged younger generations to pursue education as a means of attaining better economic opportunities outside of manufacturing or agricultural work. These changes also reflected the Black community's complex relationship to agricultural labor, particularly the history of exploitation of Black skills and labor during and after slavery, especially in the American South.[22] By the 1980s, young people growing up in New Orleans were not seeing or hearing about the kind of prolific urban cultivation practice that had once been a common sight in the city.

But there were four notable exceptions to this period of diminishing urban cultivation practice during the second half of the twentieth century: a horticultural educational program at local high schools, an urban farm at a housing project, market farming in a Vietnamese ethnic enclave in New Orleans East, and the community garden movement. The first three were successful efforts to preserve cultural heritage while capitalizing on emergent economic opportunities, and these efforts were championed by skilled growers who produced food at scale, shared their harvest with the community, and tried to pass on the knowledge. The establishment and sustenance of these cultivation projects were facilitated by the support of state or local government or local non-government organizations in securing land and funding. But much of this practice remained unknown to most of New Orleanians outside the immediate communities involved. By contrast, the fourth case—the community garden movement—was initiated by a local nonprofit organization to reactivate vacant lots by engaging new gardeners. This effort received more public attention from the media for its initial success, but not all community gardens lasted beyond their first decade.

Opened in 1942, Booker T. Washington was the first high school built specifically for the city's Black students, and it offered several vocational programs, including agricultural training. Beginning in 1981, the agriculture program expanded significantly as the "Cooperative Agriculture and Education Program" under the direction of instructor Floyd Jenkins.[23] The program eventually expanded to include an aquaculture system that yielded sufficient fish for sale and donation. One *Times-Picayune* report on Booker T. Washington High School's program quoted Mr. Jenkins, who grew up on a farm, describing how urbanites "never think of where their food came from before buying it at a grocery. Every person, especially urban persons, should take an agriculture class to know the importance of agriculture."[24] The program also partnered with George Washington Carver Senior High School's agricultural program, and was active through 2005. But the flooding after Hurricane Katrina caused severe structural damage to both schools, leading to the discontinuation of these programs. The state-run Recovery School District that took over the management of local schools after Katrina eventually decided to demolish Booker T. Washington high school buildings, which had been on the National Register of Historic Places. The new KIPP charter high school bearing the Booker T. Washington name no longer offers an agricultural program.

The Lafitte Housing Project in the Sixth Ward had an active farm on site that was set up by the federally funded Urban Gardening Program in the early 1980s. This was the "one urban farm" that the nonprofit director mentioned in the bike-tour article quoted at the beginning of this chapter. With the support of the Orleans Parish office of the Louisiana Cooperative Extension Service and Parkway Partners, a group of elderly African American men from the neighborhood tended the farm. A historical report on Lafitte Housing Project by the Creole Genealogical and Historical Association describes the garden as follows:

> Each apartment building had wire fencing and the structures occupied three/fourths of the land while the rest was landscaped to provide play

space and family gardens. Even today, many older citizens speak of the beautiful gardens that once occupied the grounds of the Lafitte and the pride that families took in their homes.[25]

The director of a local nonprofit organization supporting greening and gardening practices recalled the garden that was an entire city block as a "club" for "elderly African American men" who grew what they liked to grow, especially greens, and would give the food away to community members. But a combination of many elderly growers' aging out, waves of post-Katrina relocation, and the demolition of the housing complex in 2008 led to the end of this unique urban farming project in the city.

The Versailles Villages was established in New Orleans East in 1975 as a settlement for Vietnamese refugees. Shortly after their arrival, the residents of this new enclave began growing food in the spaces behind the apartment complex. While the practice was initially penalized and regulated because of its expansion beyond the complex's property lines, the Catholic Charities and the Mary Queen of Vietnam Catholic Church worked to legally permit the use of the land for cultivation.[26] The growers brought agricultural expertise from their rural backgrounds in Vietnam, and growing food served as a supplement for food security, and also as a site for both leisure and healing, especially for the elderly residents.[27] These growers established a street market in the early morning on weekends, where they sold produce, mostly Southeast Asian staple vegetables and livestock, alongside other value-added goods on the ground or on tables in a parking lot, similarly to the way street markets operate in many parts of East and Southeast Asia. When a large plot of land by Bayou Pratt was made available to dozens of these growers, the scale of production expanded, and the community was able to sell their excess at the markets and to local restaurants. But these markets and commercial agricultural production were not legally sanctioned at the time, and remained mostly unknown to people outside of the community through 2005.

Parkway Partner was established in 1982 to maintain the city's green spaces, especially the street medians, or the "neutral grounds" as they are called in New Orleans, as a form of private outsourcing of the city's public space maintenance work in the face of declining budget. In the early 1990s, spearheaded by Kris Pottharst, Parkway Partners began promoting community gardening in the city, especially on the lots that owed back taxes and were under city control.[28] The concept of community gardening was new to many New Orleanians at the time, but it took off quickly once the residents recognized its value. The gardens were established throughout the city, facilitated by the program that worked to identify committed, prospective gardeners and sufficient community buy-in before setting up new garden sites.[29] As these sites grew in importance for their communities, members used the gardens to gather for social events such as children's birthday parties or small weddings. At its peak, Parkway Partners counted over one hundred community gardens and school gardens throughout the city.[30] The initiative's success was recognized locally and nationally. But the number began to dwindle, and there were less than thirty active gardens remaining by 2003, according to the organization's records.[31] Similarly to the Lafitte Housing garden, people with close knowledge of the fates of these community gardens attributed their waning primarily to generational turnover. For many, when elderly gardeners stopped tending when they moved, passed away, or could no longer physically manage the work, the next generation had not joined the waitlist to enter the gardens. Other gardens were lost to the development when real estate speculation convinced owners to sell the property, because Parkway Partners did not themselves own these spaces.

Cycles of Urban Cultivation in New Orleans

The historic cycle of the popularity of gardening and farming in the city indicates that a confluence of personal and structural factors created a unique set of opportunities for growers during times of distress. The pre-Katrina trajectory of urban cultivation in New Orleans reveals predictable patterns of when and how urban gardening may take hold in the city. First, opportunities and needs for urban cultivation must be understood in the context of the city's political-economic conditions. The economic cycles of boom and bust, changing population size and composition, and shifting cultural and social environments created fluctuations in the availability of land, the urgency of urban food production, and the resources available for cultivation, especially for marginalized populations. During each cycle, land vacancy and food insecurity have been key factors in motivating urban cultivation to emerge in the city. But that is just a part of the story.

To be sure, external opportunities were not necessary preconditions for urban cultivation to thrive. For example, the abundance of vacant and blighted properties in the city starting around 1970s was a consequence of post-integration white flight, the oil industry bust, and the crack epidemic. Blight management and inspiration from other American cities prompted the community garden movement, but not until the 1980s, trailing the national trend.[32] Even then, the movement's popularity did not last for more than a decade, despite the persistent problems of vacancy and food insecurity. Economic strain and land vacancy do not automatically encourage or sustain gardening or farming in the city. Institutional resources could turn these conditions into an opportunity for urban cultivation, as demonstrated by the successes of the locally organized community garden movement and the federally funded war gardens, but only so long as the support is sustained. These efforts tended to align with the selective valorization of urban gardening as an effective, temporary measure for overcoming times of economic and social distress, rather than a recognition of gardening as a vital part of urban life. Often, these external factors stood in contrast to the personal and social reasons why the gardeners were engaged in urban cultivation.

Further, the gardeners' interest in urban cultivation was not simply reactionary, nor was it inherent. There were many reasons why people grew food in the city. It was both out of necessity and for pleasure, for themselves and for the community. Gardening appealed to people at different times for different reasons. For the recently migrated Black and immigrant population with limited economic opportunities in the early twentieth century, growing food in the city was a form of collective efficacy (e.g., community food security), entrepreneurialism (e.g., truck farming and street markets), and cultural preservation. Over time, for many reasons, these practices of local, communal food provisioning ceased to be part of the way of life in these communities.[33] One possibility is that the painful memories associated with slavery and sharecropping diminished interest in engagement with gardening and farming among younger generations in Black communities. Desegregation of commerce and educational opportunities, along with modern conveniences such as processed food, could have enticed Black consumers to participate in the new foodways of shopping at grocery stores rather than growing their own food. The life-stage cycle also affects who can engage in this type of cultivation. Older people are most likely to have time to engage in daily garden maintenance, but the physically demanding nature of the work creates challenges for this demographic group over time. These personal factors are just as important to take into consideration when understanding the cycles of urban cultivation popularity. In short, it is not the case that some people are simply "born with the green thumb" and thus would take up gardening anytime, anywhere. Moreover, we learn from these histories that people's interest in gardening did not result in sustained practice, as seen in the cases of declining truck-farming practices and the decline of victory gardens.

This brief overview illustrates the range and forms of urban cultivation that predated Hurricanes Katrina and Rita, and where things were in the city when the storms and resultant flooding decimated eighty percent of the city's housing and infrastructure and scattered eighty percent of the city's population across the country, some of them only temporarily and others permanently. It shows that growing and procuring food had long been a part of life in the city, but why and how these gardens came about and who were growing varied across time. The rest of this chapter will pick up on this timeline and document in more detail how changes in the city shaped and guided the development of the new urban cultivation scene over the decade following the 2005 disaster.

After the Storm: The Post-Katrina Emergence of Urban Cultivation

The emergence of urban cultivation in the city after 2005 occurred specifically in the context of the city's transition from recovery to redevelopment. Proliferation and increased visibility of urban cultivation in New Orleans during the decade following the storms reflected both demographic and economic changes in the city. By the summer of 2015, there were over seventy full-time or nearly full-time growers in the city tending nearly ninety lots, not including community gardens (see Tables 1.1 and 1.2). Some growers were operating in small backyard lots, while others cultivated in spaces spanning a dozen lots. Urban cultivation projects were located throughout the city, though there was some

TABLE 1.1. Number of Urban Cultivation Sites by Lots and Type in New
Orleans 2005–2015, Data collected by author from personal interviews and
the US Census.

Garden Type	2005	2006	2007	2008	2009	2010	2011	2012	2013	2014	2015
Community Garden	26	19	19	26	32	34	27	40	46	40	43
Urban Garden and Farm	1	2	3	7	24	32	37	52	67	85	88

notable concentration of projects in areas that were heavily flooded in
2005 and in predominantly African American areas (see Figures 1.1–3).[34]

While the growers' demographics remained fairly diverse, by 2012,
the majority of the growers operating in Orleans Parish were under the
age of 35, white, and had moved to the city since 2005 (see Table 1.2).[35] It
is important to note that some *novice* growers, meaning those who had
only recently taken up growing[36] as full-time or primary occupation,
were older than 50; their number was larger than those of similar age
who had been growing in New Orleans since before 2005.

In the subsequent sections, I document the economic, political, and
social changes in New Orleans over the decade following the storms
had direct and indirect impacts on the development of urban cultivation practices. I have divided the decade into three approximate time
segments: 2005–2008 (recovery), 2009–2011 (transitional), and 2012–
2015 (redevelopment). The segmenting of the timeline is not meant to
indicate that a particular segment had a distinctive beginning and ending, as some may argue that parts of the city are in a perpetuate state of
recovery, since the disaster and the pace of the recovery-to-development
transition varied significantly across the city. Instead, the timeline here
is intended to distinguish each period based on the dominant framing
of the city's status by the government and the media, which both guided
and reflected the types of policies and economic decisions that were
being made during these periods.

Urban cultivation in post-Katrina New Orleans began in a city that
was not ready for or fully supportive of the idea, despite the projects'
focus on recovery and community redevelopment. As the city began to

Figure 1.1. 2008 New Orleans Urban Farm Locations and percent Black African American population by 2000 Census Tract. Map by author using the GIS software QGIS.

Figure 1.2. 2012 New Orleans Urban Farm Locations and percent Black African American population by 2010 Census Tract. Map by author using the GIS software QGIS.

Figure 1.3. 2015 New Orleans Urban Farm Locations and percent Black African American population by 2015 Census Tract. Map by author using the GIS software QGIS.

TABLE 1.2. New Orleans Grower Demographic Characteristics 2007–2015, Data collected by author from personal interviews and archival research.

	2008	2012	2015
Arrival in New Orleans			
Born or has family roots in New Orleans	5	12	12
Arrived prior to Hurricane Katrina	8	14	16
Arrived following Hurricane Katrina	3	34	43
Total	16	60	71
Age			
20–29	4	19	15
30–39	4	9	17
40–49	1	2	11
50–59	5	6	4
60 and over	2	2	5
Total	16	38	52
Gender			
Male	8	22	31
Female	8	26	28
Other	0	0	2
Total	16	48	61
Race			
Black/African American	9	12	12
White/Caucasian	7	40	49
Asian	0	7	9
Other	0	1	1
Total	16	60	71

shift its priorities from recovery to redevelopment, the interests and support for the cultivation projects began to coalesce, even as some parts of the city struggled to return and rebuild. During the last few years of the decade, conditions in the city ripened for commercial urban cultivation: the new food economy and tourism were in full swing, and there was increased in-migration of young, middle-class people into the newly gentrifying neighborhoods. The following timeline will serve as an important backdrop for the remainder of the book, in which we learn more about what made the growers decide to pursue urban cultivation (Chapter Two), how they managed to get the projects up and running (Chapter Three), and why some of them succeeded in turning their visions into reality, while others struggled (Chapters Four and Five).

Post-Disaster Recovery Period (2005–2008): Seeds Are Sown

Post-disaster New Orleans was not necessarily the most fertile environment for urban cultivation. While a number of community gardens continued or resumed their operations, the type of urban cultivation projects on which this book focuses was slow to emerge during this period. Few new forms of gardens and farms had begun to operate in the city by 2008, but they were not a part of the city's recovery agenda, and most of the projects developed independently across the city, without a coordinated attempt to systematically integrate urban cultivation into the recovery efforts at large.

The major flooding and destruction created conditions that some long-term resident and New Orleans-native growers later described as a "wake-up call" regarding food insecurity and ecological vulnerability in the city. When nearly all grocery stores were closed or operated at limited capacity in the months following the storms, residents who returned to the city faced an acute sense of food insecurity, regardless of their economic means. The ecological vulnerability of the city to the flooding, a result of the combined conditions of coastal erosion, poor engineering of the levees, and climate change, elevated the residents' concerns about sustainability in anticipation of future storms. Even in the areas that did not take as much water, the soil and air became exposed to toxic chemicals as a result of flooding.[37] These concerns led to a contentious debate over the future of the city: should it be rebuilt, especially in the areas that were at high risk of flooding?

Nothing embodied the tension over whether to rebuild the city more acutely than the controversy over the "Green Dot" map. The *Times-Picayune* published the infamous map in January 2006. The map, which was a reproduction of the "Action Plan for New Orleans" report by the Bring New Orleans Back Commission Urban Planning Committee, marked the areas that experienced severe flooding with large green dots. The six green circles in the original map presented by the Urban Institute were identified in the legend as the "areas for future parkland." This was one of the recommendations by the committee and not a formal policy, but the public outcry was immediate and intense, and the map, along with the urban shrinkage plan, disappeared before it was ever seriously considered.[38] Only one of the green dots was placed in a middle-class neighborhood, Broadmoor, and the rest were placed on predominantly working-class and Black neighborhoods. This incident imprinted in the minds of many long-term Black residents the language of *green spaces* or *greening* as the symbolic manifestation of the city's attempts to prevent their communities from returning, years before scholars began coining phrases like "ecological gentrification" or "green gentrification."

Blight became a renewed challenge for New Orleans when Hurricanes Katrina and Rita and the resulting flooding destroyed more than 800,000 homes across the city. The city had already been struggling with the high rate of blight, with 26,000 properties classified as blighted in 2005, before the flood.[39] But in the months and years following the storms, many more abandoned lots were taken over by weeds and unruly vegetation, and became a dumping ground for trash, cars, chemicals, and even dead animals, as the property owners struggled to rebuild. By one estimate, the city had over 70,000 blighted or otherwise vacant properties in 2008.[40] Before the storms, New Orleans had a relatively high rate of homeownership among low-income families, especially in predominantly Black neighborhoods. According to the 2000 Census, 57 percent of the occupied housing units were owned in the Lower Ninth Ward, where 97 percent of the residents identified as Non-Hispanic Black or African American.[41] Just over half of the Black residents were able to return to New Orleans fourteen months after the storm, compared to more than 70 percent of their white counterparts.[42] The uneven rates of residents' return to the city created spatial concentrations of blighted properties in predominantly Black neighborhoods.

Black homeowners faced a multitude of discriminatory hurdles in their efforts to return and rebuild, especially from a bureaucracy that produced disproportionate implicit bias in favoring white homeowners over Black owners and renters when it came to assisting in the recovery and resettlement process. The Louisiana Road Home program was supposed to offer the state's homeowners financial compensation to expedite the rebuilding process of their homes or help them relocate.

Yet by design, the program penalized those who owned in areas where property values had depreciated or stagnated.[43] Residents who had flood or homeowner's insurance had to endure lengthy battles with insurance companies whose responses were inconsistent, opaque, and inadequate. Very often the documents needed to apply for these programs had been lost in the muck of mud, in the mail, or at government offices. If the homes were passed down from parents or grandparents, the current occupants might not have known of the physical deed's location or had flood insurance policies. Louisiana's "intestate succession" law[44] has resulted in "heir's property" problems that disproportionately affected African American owners, contributing to their individual and collective loss of assets as it took them longer to rebuild their homes and communities or to sell their property to recuperate its value.[45]

Some areas and issues received more attention from the media. The Lower Ninth Ward, much of which remained under more than eight feet of water for weeks, received disproportionate media attention, even though other areas were similarly devastated (e.g., New Orleans East, Lakeview, and Gentilly). Its newly established status as the disaster's most prominent scar spurred curiosity among outsiders, who *flooded* the neighborhood as volunteers for recovery work and later as disaster tourists. Other neighborhoods received far less attention from the national and international media for their ongoing struggle for rescue, recovery, and rebuilding efforts. Such skewed attention to the Lower Ninth Ward was satirized in an episode of David Simon's *Tremé*, a fictional television series about post-Katrina New Orleans, in which a musician confronts a group of young volunteers from Wisconsin that expressed empathy and disbelief toward the slow recovery of the "Ninth Ward" by asking them, "Let me ask you something. You even heard of the Ninth Ward before the storm? Well . . . So why are you so fired up about it now?"[46]

Even though an extraordinary amount of attention and resources were brought to the Lower Ninth Ward, the neighborhood's recovery rates lagged behind many other neighborhoods that received little or no external attention and support. A large portion of the Lower Ninth Ward remained vacant over the decade.[47] By contrast, the middle-class Broadmoor neighborhood rebuilt relatively quickly, with a large number of the residents privately funding the rebuilding, while Central City and

Mid-City eventually saw rebuilding and repopulation partly due to the new developments funded by public-private partnerships.

During the first few years after the storms, the number of nonprofit organizations servicing the city, especially in the Lower Ninth Ward, proliferated, albeit with limited success in rebuilding homes and communities.[48] During the recovery period, hundreds of thousands of volunteers arrived in New Orleans each year. Some came for a short period of time to assist in gutting houses or cleaning up the debris, while others stayed for an extended period of time to work as a part of AmeriCorps or Teach for America. Many were younger, college-educated, and were described by some as Young Urban Rebuilding Professionals (YURPs).[49] Some of these volunteers and service workers made New Orleans home after their initial service was completed, and they sought continued opportunities to engage with the recovery process.

In this climate, a handful of the new types of gardens and farms began alongside the few community gardens that remained from before the storms. These new projects were tended by the long-term growers. Some of the new gardens were associated with one of many nonprofit organizations primarily focused on recovery work or started by those who arrived in the city to participate in the recovery efforts. For example, the Common Ground Relief's bioremediation gardens were in operation in the Lower Ninth Ward by 2006, and the first Edible School Yard in New Orleans was established in 2006 as part of one of the new charter schools in the city, modeled after a program initiated by restauranter Alice Waters in Berkeley, California. Other gardens were started by long-term residents or people with roots in the city as a part of their community rebuilding efforts.

Urban cultivation was not a part of the dominant vision for the recovery at the time. In most cases, the idea of using a garden as a part of community rebuilding did not readily resonate with the returning residents, whose focus at the time was on housing, schools, and other major infrastructure. According to Parkway Partners' census of the community gardens in fall 2005, there were only 13 gardens already operating in the city, with 29 described as being ready or having potential "to be activated with community interest."[50] Public policy largely remained uninterested in urban cultivation, even as city and state governments dealt with the growing number of blighted properties and the mounting cost of maintaining the Road Home properties sold to the state. If the governments were not promoting gardens and farms as a part of the post-disaster recovery agenda, they were also not regulating or facilitating them. Until 2014, only community gardening was a legally permitted form of cultivation in New Orleans aside from home gardens. These new forms of single-grower gardens did not fit the zoning code, but none of the growers reported being penalized for their supposed violations during this time. Overall, the recovery period remained an exploratory and preparatory era for new urban cultivation projects. By 2008, a handful of urban cultivation projects found some footing in their operations.

Transitional Period (2009–2011): Ideas Sprout

The number of non-community garden sites being cultivated tripled from 2008 to 2009, signaling a sudden expansion of the practice in the city. As the city began to shift priorities from recovery to redevelopment, urban gardening began to resonate more readily with the visions of the city's future held by both developers and community organizers. On the one hand, in the areas of the city that recovered quickly, urban cultivation could be seen as a sign of rebirth, that the city had not just come back but was exhibiting new strength by converting gray and brown to green. On the other hand, in the areas that continued to struggle to recover, the realization settled in that neighbors might never return or build on vacant properties. This made the idea of leasing the lots to growers more appealing to the landowners, both public and private. The social and organizational landscapes of the city shifted drastically during this period, creating new opportunities and support systems for the cultivation projects. In this context, more gardens began to take root across the city, and most of the earlier urban cultivation projects evolved and expanded to take advantage of the new opportunities.

The New Orleans Saints' Superbowl victory in 2010 symbolized optimism for the city's comeback. As the team was welcomed home with a characteristically extravagant parade with Mardi Gras floats, the city seemed ready to transition from the post-disaster recovery mode toward economic, political, and demographic stabilization and growth. Tourism rebounded quickly and became an even stronger economic driver, thanks to the city's prioritizing of rebuilding tourism infrastructure, creating tax incentives, and extensive marketing of an "authentic" New Orleans.[51] By 2010, New Orleans seemed "open to business" to many visitors who strolled through the French Quarter or the Garden District and enjoyed meals at one of the city's many fine dining establishments. In fast-recovering neighborhoods, such as Uptown, Carrollton, Broadmoor, and Bywater, most damaged homes were rebuilt, blighted properties were razed over, new charter schools opened, and businesses operated on regular schedules.

By contrast, other areas remained underpopulated, with properties blighted or overgrown and businesses slow to reopen or enter these areas. In May 2009, FEMA announced that residents still living in federally-issued trailers must vacate, regardless of whether their homes had been completely rebuilt. The Lower Ninth Ward neighborhood's recovery remained frustratingly slow, except for the well-publicized "Make It Right houses." Around 2008, the nonprofit organization, founded by Brad Pitt, began building dozens of homes along Tennessee Street, right where the industrial canal levee had breached. The houses were designed by imaginative architects and engineers, who claimed that they were sustainable and could withstand future disasters. The modern architectural style of these homes stood out amid the neighborhood's traditional shotgun architecture, but they were promoted as a beacon of hope for the *new*, sustainable rebirth of the city.[52] The Make It Right houses created a stark juxtaposition with the remnants of destruction still visible in the neighborhood, and this area became the most troubling destination for disaster tourism.[53] Tour buses and vans rolled down the streets of the Lower Ninth Ward, occasionally making stops for visitors to take photos of the dilapidated buildings or overgrown lots few blocks away from these houses, often without ever leaving the bus. Some angry residents posted signs calling out the insensitivity and the profiteering nature of these tours, with one sign stating, "TOURIST, SHAME ON YOU DRIVING BY WITHOUT STOPPING, PAYING TO SEE MY PAIN, 1,600 DIED HERE." There were several urban farms only a few blocks away from the "Brad Pitt houses," but the tours left the neighborhood without noticing these cultivation sites.

As the grocery stores in recovered areas returned to the normal operations, the food access disparity in the city reflected the geography of race and class injustice in the city. The opening of two new alternative food markets were part of an effort to address persistent food inse-curity in some of the neighborhoods that were struggling to rebuild. Hollygrove Market and Farm opened in October 2008 in the Holly-grove neighborhood, where the rapper Lil Wayne grew up. The major-ity of the neighborhood took over six feet of water during the 2005 flooding, and recovery in the area was slow. Carrollton-Hollygrove Community Development Corporation, along with a few partnering local nonprofit organizations, had established a small market on a former nursery site that did not reopen after the storms. Though the idea for the market and the gardening predated the storms, the new-found availability of space and the renewed interests in addressing food insecurity led to the opening of the new market. The market aggregated and distributed locally grown produce, initially starting with a $25 produce box, similar to the community-supported agri-culture (CSA) model, but without requiring membership or advance payment. It started with weekly hours on Saturdays, and by the sum-mer of 2011, it had added Tuesday afternoon hours, due to increased demand. The market bought small amounts of produce without con-tractual commitments, which created an opportunity for urban grow-ers to sell their vegetables and herbs to the market, though most of the produce sold at the market came from rural producers in Louisiana, Mississippi, and Alabama.

Across town in the Marigny neighborhood, near the French Quarter, New Orleans Food Co-op opened in 2011. There had been a plan to open a brick-and-mortar business at another location in Marigny in summer 2005, but the storms shattered that original plan. The co-op eventually found a new location inside the New Orleans Healing Center, across from the St. Roch Market that remained blighted since the storms for nearly a decade until it would eventually reopen as a high-end food hall in 2015. The co-op was open to the public for shopping and emphasized organic and regional food products, though the emphasis was not on food grown within the city. Both of the new alternative markets were located in areas that had been identified as "food deserts"[54] in hopes of increasing access to healthier food options within these neighborhoods. Nevertheless, the success of these markets was associated with a cus-tomer base of younger, middle-class newcomers who continued to arrive to work in recovery or to enter the city's slowly recovering economy.[55]

Unlike the recovery period, when gardening projects were operating mostly independently and without wider recognition across the city, the organizational support for urban cultivation expanded during the transitional period, especially for full-time growers. Philanthropic organizations such as the Greater New Orleans Foundation and the Kellogg Foundation recognized the potential of gardens and farms in the city to address public health and youth education; this in turn allowed growers to access funds to continue or scale up their projects. Younger people affiliated with service organizations such as AmeriCorps and Teach for America came to New Orleans to fulfill their year or two of service, but unlike the YURPs that arrived during the recovery period, much of their work focused on building long-term resilience in the form of environmental sustainability or educational equity. A couple of the AmeriCorps members working in New Orleans during this period were paired with nonprofit organizations focused on urban cultivation, including Hollygrove Market and Farm and Our School at Blair Grocery. Local universities began to provide support to the emerging urban cultivation projects by connecting growers to financial and non-financial resources. For example, Tulane University's City Center[56] worked with several urban cultivation projects, including Hollygrove Market and Farm and Grow Dat Youth Farm, a youth leadership program, by designing and building a physical structure onsite for each. Faculty members from the University of New Orleans also offered their expertise in landscape design to a few other cultivation projects.

Local nonprofit organizations that were previously focused on community gardening or backyard gardening began to provide resources and services to this new type of cultivation project, with an increasing attention to the commercial potential of growing food within the city. Parkway Partners, which was instrumental in the earlier wave of community gardening in the city during the 1990s, had reassessed and reactivated some of the former community gardens during the recovery period. Some of the growers at these gardens, mostly new to the city, began to explore the potential of scaling up their cultivation or entering commercial production by 2011. Parkway Partners offered free seeds and workshops by master gardeners to novice growers. The growers had access to its greenhouse, the only publicly accessible one in town at the time, making it possible to start seedlings ahead of the planting season. These resources were instrumental for many new growers to develop interest in and capacity for full-time cultivation.

The New Orleans Food and Farm Network (NOFFN) was established in 2002 to assist and promote sustainable growing practices in the city, primarily focused on the health and environmental benefits of backyard gardening. After the storms, the organization continued to advocate for local gardeners and growers, while also creating the neighborhood-based NOLA Food Maps program to facilitate food access. The organization began to focus more explicitly on urban farming, especially market growers, after Sanjay Kharod became the director in 2010. It began offering more workshops focused on land access, grant applications, and financial management of urban farming, and notably catered to the needs of full-time growers rather than hobby gardeners.

Both Parkway Partners and NOFFN served as umbrella organizations for grant applications for urban cultivation projects that required non-profit tax status and liability insurance. There were some ambiguities in terms of how the two organizations' missions and practical territories were designated, as they responded to the changing urban cultivation scene and the broader social, economic, and political changes in the city. But the growers by and large were not expected to pledge allegiance to one or the other, and nearly all spoke graciously about the services that these nonprofit organizations offered.

The increasing cultivation activities in the city and its land-management potential did not escape the attention of local and the state governments, as they began to take a more aggressive approach in re-introducing these properties into the private market. The New Orleans Redevelopment Authority (NORA), a state agency that had been tasked to reintroduce the Road Home properties to the private market through commercial development and private sales, established the Lot Next Door program in 2007. The program gave the neighboring property owners, whose lots shared the front property line with a Road Home lot, the first rights of refusal for purchase. Seeing a tepid response to the program, in 2009, NORA introduced Growing Home, a sub-program of Lot Next Door, which allowed Lot Next Door participants to allocate up to $10,000 of landscaping cost toward the purchase price of the neighboring Road Home lot if they turned it into a garden or a green space. The program was intended to make the lots affordable, but it was also designed with the hope that it would contribute to neighborhood beautification and blight prevention. This program was one of the first cases of using greening as a tactic for urban redevelopment in New Orleans since the Green Dot controversy. According to one NORA official, "nearly all" of the Lot Next Door purchases used the Growing Home program, though the majority of these "green" projects were expansions of front or back yards with some landscaping and fencing to meet the requirement.[57] None of the new urban cultivation projects discussed in this book participated in the program, since growers had to already own an adjacent property to be eligible.

Mitch Landrieu won the mayoral election in 2009, becoming the first white mayor in 40 years (as the son of the previous white mayor, Moon Landrieu), signaling political and social changes in the city that the previous mayor Ray Nagin, who is Black, infamously proclaimed in 2006 would remain a "chocolate city." The new administration began to implement measures that signaled a shift in priorities from recovery to redevelopment. Some of these changes had direct implications for urban cultivation. The Master Plan adopted by the city in 2010 urged revision of the comprehensive zoning ordinance (CZO), which had last been updated in 1995. The first public draft of the plan, released in 2011, included references to "agriculture" as a new category for land use to be permitted "in appropriate locations" where there were no development plans for the near future, in order to "increase access to healthy food at a lower environmental cost by supporting the production, processing and distribution of locally grown food."[58] The NOFFN convened a working group of stakeholders to assist in revising the section pertaining to urban agriculture in the Master Plan, and a group of growers also organized themselves separately to proactively shape the language in the new ordinance. In general, however, most growers remained nonchalant about the lack of legal protections for their practice, due to the lack of regulatory oversight.[59]

Somewhat more indirectly, the city's investment in the hospitality industry created opportunities for the cultivation projects to be aligned with these priorities in a few distinct ways. The 2009 Master Plan for the Reinvigoration of Tourism published by the New Orleans Strategic Hospitality Task Force aimed to nearly double the city's visitor count over the decade by rebranding the city to especially appeal to younger demographics.[60] In a city known for its unique and refined culinary traditions, fine dining establishments were one of the first industries to benefit from the boost in tourism. Growers successfully convinced the chefs in both well-established and newer restaurants to source a portion of their ingredients from urban producers, capitalizing on the popularity of the "farm-to-table" trend around the country. Increased tourism also meant more possible "voluntourists" who were eager to add a day of service to their vacation or business trip to New Orleans. Beyond tourism, the city began to aggressively court new businesses by creating entrepreneurial opportunities, especially in cultural sectors. The Office of Cultural Economy at City Hall promoted cultural economy expansion,[61] and the film industry began taking advantage of the Louisiana's film tax credit, resulting in a filmmaking boom in the city.[62] These measures injected vast amounts of capital into the already recovered parts of the city, enticing newcomers to visit these neighborhoods to dine, shop, film, and even consider relocating there.

Social, ecological, and economic uncertainties continued throughout this period, creating both new constraints and opportunities for the budding urban cultivation scene. Just as New Orleans' recovery seemed to have been on track for redevelopment, the Great Recession devastated the national and international economy between 2008 and 2009. Then, in April 2010, Deepwater Horizon's offshore drilling explosion became yet another environmental and economic catastrophe to devastate the Gulf Coast region, reminding the public of the vulnerability and hazard of the global political economy. Philanthropic funding for the urban cultivation projects that focused on youth reflected national concerns over childhood obesity and educational inequalities around 2008–2010. But by 2011, the economic crisis had created uncertainty about the sustainability of funding, and in some cases, foundations' priorities shifted away from obesity to other social issues such as the opioid epidemic or climate change.

Nationwide, terms such as "gig economy" and "sharing economy" came to define the new era of service economy, exemplified by Silicon Valley start-ups like Uber and Airbnb.[63] These new jobs touted flexibility and independence as the *perks* of being one's own boss, while masking the exploitative, unstable, and risky nature of working in these industries. The expansion of social media platforms and widespread uses of mobile devices also created crowdfunding platforms such as Kickstarter and Indiegogo. These platforms became a new way for funding one's entrepreneurial projects or social causes rather than relying on traditional banking systems or large philanthropic organizations. Though crowdfunding one's project had the appearance of independence and democracy, in reality, it reinforced existing structural-economic inequalities across racial and class lines, in terms of which projects ultimately met the funding goals and took off.[64] It also normalized continuing disinvestment from public social safety-net programs. These national trends and platforms entered New Orleans during the transitional period and began to shape how urban cultivation projects sought resources and defined their roles in the city as their practice increasingly gained recognition.

Redevelopment Period (2012–2015): Harvesting and Planning for the Next Season

In 2012, New Orleans' population increased for the first time since Katrina. But the racially disproportionate increase in population indicated that this growth reflected the entry of newcomers rather than the return of long-term residents. While Black residents remained the city's majority racial group, their proportion declined from 67.5% in 2005 to 59.8% in 2015, while the proportion of whites increased from 26.5% to 31.2% during the same period.[65] Those who transplanted after 2011 were more likely to be white, have higher levels of education, and have relocated to the city to pursue career opportunities in the emerging economy. New food establishments, from restaurants and bars to grocery stores, became notable markers of development in the neighborhoods that had recovered and were repopulating quickly. As the city neared the ten-year anniversary of Hurricane Katrina, the urban cultivation scene in New Orleans took a notable turn in terms of its proliferation, public visibility, and integration with the growing economy. The number of new urban cultivation projects increased rapidly from 37 lots in 2011 to 67 in 2013, then to nearly 90 lots in 2015. Increasing local and national media coverage of the gardens and the farms boosted their recognition and popularity, especially the newly established entrepreneurial projects.

The economic growth did not benefit everyone in the city and in fact posed new threats to the homeowners and communities that continued to struggle to rebuild. A *New York Times Magazine* article in 2012 described the Lower Ninth Ward as "Jungleland," describing the neighborhood as being "reverse colonized" by the forces of nature.[66] Local activists criticized the article's description as perpetuating negative stereotypes of the neighborhood and overlooking the ongoing efforts within the community to rebuild itself beyond simply cutting grass to keep the overgrowth under control, which was the single form of community-led effort mentioned in the article.[67] Nevertheless, vacancy and blight had become the major redevelopment priorities of the Landrieu administration's first term. The city began aggressively enforcing blight violation and demolition of the lots that failed to comply with the existing ordinance and proclaimed a plan to reduce blight by 10,000 lots by 2014. The emphasis of policy enforcement was placed on clearing the lots or reintroducing them to the market, without a clear agenda on what should be built on these cleared lots, especially in the areas where post-disaster repopulation had been slow. The city never established a formal urban agriculture policy to use gardens and farms as a means of blight remediation or vacancy placeholders, but other organizations began to see urban cultivation as a viable solution: someone would pay to lease the lot and maintain it as a cultivation site.

The most explicit attempt to co-opt urban cultivation was a series of programs that NORA implemented between 2013 and 2015. Unlike its earlier programs, the new programs explicitly courted urban growers to use the lot for urban agriculture. It began with Lot Next Door 2.0 in 2013, followed by Lot Next Door 3.0 in 2014, each program expanding the potential pool of eligible applicants to purchase the Road Home properties, first by changing the property lines that could be shared and then removing the homestead requirement altogether. It introduced the Growing Green initiative in 2014, a leasing program for anyone who wish to use a blighted lot for *green* uses, including rain gardens, tree shading, and urban farms. The agency made no secret of its general approach to these programs as a good *temporary* use of currently underutilized or undervalued lots, even though Growing Green offered the program participants a possibility of purchasing a lot after successful lease partnership during the first few years.[68]

Nonprofit organizations and private lot owners also reached out to the growers to lease their lots for cultivation. Habitat for Humanity, which played a key role in rebuilding homes in the region after the storms, initiated the Habitat Urban Garden (HUG) Fresh Food Initiative in January 2012. At the rate of $1 per year, HUG leased donated lots that were not ready for building homes due to their location, size, and configuration. As of 2015 it was leasing 15 lots to full-time growers. Private landowners became more concerned with maintaining their vacant lots to avoid blight violation fines due to vegetation overgrowth. Some of these owners could not yet rebuild on their lots, while others owned the properties and did not see the immediate financial need or potential to build on those lots. Not yet ready to sell their properties, which were often family inheritances, these private lot owners found value in urban cultivation not just for maintaining their lots but also to support the growers' work in the community as they got to know their work during the transitional period.

In addition to land access, the potential for commercial production expanded during this period when the city saw an explosion of a new food economy that was closely connected to gentrification.[69] The upscale mixologist bar CURE opened on Freret Street in 2011, initiating a series of new businesses opening within the surrounding blocks over the following four years. It was one of the most recognizable cases of a food establishment activating the commercial potential of the areas of the city that had long experienced economic decline.[70] Across the city, new restaurants seemed to open each week; most featured food items that were a "twist on" or "reimagined versions" of traditional New Orleans cuisine, or that were a New Orleanization of American classics such as hot dogs and burgers. These new venues marketed what Cate Irvin calls a "hybridized authenticity" of the food and the experiential aspects of the emerging new food economy.[71] Geographer and activist Jeanne Firth documented the emergence of local chefs as celebrity philanthropists during this period whose profile and influence reached beyond the restaurant industry.[72] Facilitated by the increasing use of social media platforms like Facebook and Twitter among young and middle-class consumers, especially those who described themselves as "foodies," the new food economy seemed to embody the city's rebirth as a new place with new people who now called it home, while quietly looking past those who were economically and socially excluded from the opportunities in the *new* New Orleans.

Urban growers not only took advantage of the increased demands for locally grown produce but also actively engaged with the increased interests in their practice by marketing themselves through collaboration with other businesses or initiatives that started around this time. The garden spaces became event venues for pop-up experiences, temporary retail set-ups for businesses without a permanent brick-and-mortar space, or fundraising sites for nonprofit organizations. These events often featured some of the newer, *fine-casual* dining restaurants' offerings paired with the drinks from a number of new local breweries or distilleries. The first Eat Local Challenge in New Orleans took place in June 2011; it encouraged participants to commit to consuming only food "grown, caught or raised within 200 miles of New Orleans" for 30 days.[73] These events and the subsequent media coverage of associated events, such as the urban farm bike tour described at the beginning of the chapter, helped raise the profile of urban cultivation in the city.

For the entrepreneurial cultivation projects, an expanding market and increased demand were not the only new opportunities. Propeller, a local incubator for social ventures, was established in 2009 to provide funding and mentorship for emerging small businesses and nonprofit organizations. Propeller launched PitchNOLA in 2011, a competition for projects that offered cash prizes. Its three primary foci—"community solutions," "lots of progress," and "living well"—encouraged a number of "innovative" urban cultivation projects to throw their hat in the ring. The judges included local business professionals and academics, and the participants "pitched" their ideas in a public forum akin to the reality TV competition series *Shark Tank* for funding. Eight urban cultivation projects participated in the PitchNOLA contest or became accelerator fellows between 2011 and 2015, including both nonprofit and for-profit projects.

In May 2015, the city council passed the revised Comprehensive Zoning Ordinance (CZO), which went into effect on August 15 of that year, just two weeks before the tenth anniversary of the Katrina landfall. With it, "agriculture" became a legal land-use category, but with it also came newly formalized regulatory policies. The adopted CZO's Agriculture section (Article 20. 3.C) contained regulations that were not congruent with the practices of urban cultivation in New Orleans at the time and did not reflect the suggested language proposed by the growers and the NOFFN. For example, it included an extensive list of chemicals and hard metals for which soil testing was required, even though there was no lab in the state that could conduct such testing, including Louisiana State University's Ag Center. The new CZO also limited processing in residentially zoned spaces only on a conditional basis, disregarding the fact that most of the cultivation projects were not producing at the scale that would justify the use of the industrially zoned facilities for processing, such as washing and bundling the produce. In reality, however, the changes in the legal status of and restrictions on urban agriculture did not change growers' practices, and they continued to grow, process, and sell or donate their produce to restaurants, markets, and directly to individual consumers as they had already been doing.

By 2015, it seemed as though urban cultivation had finally found its maturity in the city. But the new economic opportunities also presented some downsides. For example, a San Francisco–based online farmers market venture called Good Eggs opened a retail store in New Orleans in 2013 and quickly signed up many local urban and rural growers as its providers. But in less than two years it abruptly closed the retail location, stranding consumers and producers who had shifted their production and distribution to work with the business. Land tenure now became a major concern for growers, especially those operating in rapidly gentrifying neighborhoods, to an extent that it had not been during the recovery or transitional periods, including in the Lower Ninth Ward.[74] The growers were also recognizing that operating urban cultivation projects, whether as nonprofits or for-profits, was not economically sustainable for most of them, despite the projects' seeming popularity and success. Regardless of access to land and resources, there were signs of burnout or restlessness to reach the next stage in their lives, especially among some of the younger growers, many of whom began their cultivation projects in their late twenties and early thirties.

Conclusion

There has always been a connection between gardening and urban disasters, acute or perpetual, and as we've seen, different forces seem to give rise to particular forms of urban cultivation in reaction to different types of disasters. Folk urban cultivation in the city's Black communities from the late nineteenth century to early part of the twentieth century reflected the racial capitalism that exploited Black labor in the agricultural industry, first via slavery and then via sharecropping, and enforced disparate access to resources along racial and class lines in the city. Growing food for the community used to be a form of collective efficacy in response to structural injustice. War garden and community garden programs were programs implemented and funded by state agencies or NGOs to boost patriotism, teach vocational skillsets, or spur neighborhood revitalization. The urban cultivation scene that emerged in the aftermath of the 2005 flooding of the city did not follow either of the previous iterations of urban cultivation practices. It developed outside of the community tradition, grassroots activism, and top-down recovery agenda, and manifested as an impromptu network of individualized actions that developed quickly in correspondence with the significant changes taking place in the city.

This chapter's review of the post-disaster transformation of the city illustrates that the emergence of urban cultivation in post-Katrina New Orleans took place during a time of great uncertainty and rapid transformation. The growers were enacting their visions of alternative food systems, land use, and lifestyles in the city as the conditions around them continued to shift, posing significant challenges while also creating new opportunities for their practice and new interest from others. Readers may return to this chapter's overview of the post-Katrina decade in New Orleans as they proceed through the rest of the book, in which I will unpack in more detail the relationship between urban cultivation and post-disaster urban transformation as a form of prefigurative urbanism.

2

Seeds of Hope

Imagining Alternative Ways of Living, Working, and Provisioning

"There's something that happens when people start working
with growing things and it doesn't matter who they are or
where they're from, it's just . . . something happens."
—Zach, CRISP

Expressions such as "planting the seeds," "being rooted," or "food for thought" allow us to think about gardening as a metaphor for healing, transformation, and optimism amid recovery and rebuilding efforts. Converting a vacant lot into a garden is physical evidence that change is possible. Watching the plants grow day by day, from tiny seedlings to vines and stems in a matter of weeks, reminds us of nature's power to restore and reproduce, even if we are just passing by the gardens as neighbors. The sight of children playing in a garden, picking from the plants, or chasing chickens brings a sense of normalcy and innocence, however ephemeral, to a community processing and overcoming trauma. In retrospect, it might seem obvious that urban cultivation would take off in post-Katrina New Orleans. But it was not obvious at the time, at least not immediately.

Urban cultivation has an immediate and tangible transformative quality that contrasts with the slow, complex, and unpredictable nature of bureaucratic and infrastructural post-disaster recovery processes that often take months and years,[1] and can often deprive individuals of a sense of control over the process. I argue that the growers in post-Katrina New Orleans were drawn to the elastic and expansive potential of urban cultivation as a guiding principle in their respective schemata of making sense of the world after disaster. The cultural-psychological concept of *schema* is defined as the cognitive processes of organizing thoughts and observation through interpretation, which then allows actors to set goals and identify corresponding actions.[2] Direct, prefigurative action has a unique appeal during times of disruption and uncertainty, when actors feel a heightened need to reorient their understanding of the world to determine what actions to take.[3]

Sociologist Ann Swidler theorizes culture as a micro-level, dynamic process, whereby we *use* culture to make sense of the world.[4] She argues that culture takes a more explicit, coherent form during "unsettled times" in our lives, as we aim to reorganize our "strategies of action" or develop new ones to adapt and move forward. Urban cultivation provides a range of what Swidler calls "repertoires," or specific ways of doing or viewing things that we acquire as skillsets, which gives new meaning to space, social relationships, food, and one's identity. These meanings are not inherent to urban gardens and farms, but they are ascribed or activated by those who engage in cultivation, as they try to make sense of the world and imagine the near and distant future and their own role in it through acts of planting, harvesting, and sharing food. Here, "unsettledness" refers not only to external social conditions, such as disaster, but also to personal life experiences and stages, from career changes, relationships, midlife crises, and retirement. Having a concrete vision for the immediately implementable changes *makes sense* to these individuals, because they can more specifically envision how to go from here to there. This is especially applicable for actors who seek action outside structural forms of resistance and activism, but who still wish to do more than support these efforts financially or socially from a distance. For these actors, practicing prefigurative urbanism provides a space to enact changes in ways that elicit a sense of self-efficacy.

Gardens and farms can be associated with a wide variety of outcomes, because they do not inherently align with a singular set of values or a particular political agenda. A garden can be a site of radical anarchist political organizing or a leisurely social gathering place for community elders. A farm could be a part of a nonprofit organization's youth outreach program or a commercial production site for a local entrepreneur. But most importantly, the outcomes of urban cultivation do not necessarily need to be fully articulated prior to the groundbreaking of a garden or a farm. In fact, it is possible to start a garden without knowing exactly what it is for, or why you want to garden. Its purposes and potential can continue to evolve, even as external circumstances change or as growers' interests and experiences transform over time. As we'll see, urban cultivation did not simply represent or embody hope *in general* after the storms, but it provided an opportunity for the growers to articulate and enact their willful hope through actualizing their visions for change. The heterogeneity of these motives also meant that the growers were not necessarily united or organized for a single cause, even as they appeared to be a part of the same trend and in fact often interacted with each other.

This chapter illustrates a range of aspirations that individuals and organizations initially held before they launched their cultivation projects in post-Katrina New Orleans, and how the dominant set of aspirations shifted from mostly social to largely economic outcomes over the decade. In the following sections, I will first provide snapshots of the diversity and multidimensionality of the growers' aspirations and expectations for their urban cultivation projects. Then, I describe five types of urban cultivation aspirations by highlighting how they differ in terms of the anticipated outcomes and how gardens and farms were expected to deliver these outcomes. Throughout the chapter, I demonstrate that the categories of this typology were not mutually exclusive at any given moment. To be sure, growers were never motivated by a single aspiration, and their aspiration often shifted over time as they started the work, both in response to the changing social and economic environments around them and as their capacities, knowledge, and skill grew with experience. But we will see that these aspirations did not randomly appear across the decade, indicating that while the growers individually expressed urban cultivation aspirations as emerging from their own visions, there were additional external factors that impacted who was interested in gardening and farming in the city at a given time, and what potential they saw in urban cultivation.

So Many Reasons to Grow

In post-Katrina New Orleans, growers anticipated a combination of social, economic, and personal outcomes resulting from their cultivation projects (see Table 2.1). Most urban cultivation projects were linked to multiple possible positive outcomes. For example, a young Black leader of a nonprofit organization in the Lower Ninth Ward listed multiple goals associated with the organization's new garden project in the neighborhood where she grew up:

> So, our goals with growing vegetables, fruit, edible products, are to use them as a learning platform for health education and awareness. And it's also to make fresh food more available in the community. It's also an approach to addressing blight that is just pervasive and rampant in our community. So that's one aspect of it . . . But also, our initial emphasis in doing this, our initial goal, was to grow food for our market, because we had a mobile farmers market. And if farmers didn't come to the market, we didn't have enough produce . . . And we saw the school garden as a learning platform for children to learn more about whole foods.

This is a long list of anticipated outcomes for one garden, with an emphasis on social and economic benefit. As the leader of a nonprofit organization, this individual could offer an articulated, streamlined "pitch" for the garden project during an interview with a researcher. But the range and the diversity of these outcomes echo typical responses to questions about what growers hoped to achieve through urban cultivation.

Even when the growers placed emphasis on one type of outcome, there were almost always other outcomes they foresaw. For example, Zach, who is white and arrived in New Orleans shortly before the storms, described his primary reason for starting a cultivation project in the St. Claude neighborhood as wanting to put permaculture theory into practice, something he cared deeply about. However, he also named a number of other possible positive outcomes he envisioned for the project:

> This area where you are right now, I see as more of a showcase or place for the community to come together, a place where we can have events. We've got Patrick, who actually works here a lot, and myself, who works here a lot. And it's not like we're making money, but if we could work full time to support ourselves, that would be amazing. But that can't happen when you have this lot out here, and kids are coming by who are helping, and you want them to try stuff. . . . Why I think it would be nice to grow is not so we can produce more. It's because it's going to allow other people who love this to actually make a living, to allow a kid to have a summer

job, and if nothing else, they have a job that summer, and they're going [to be] applying for college and applying somewhere and here's a reference, "He's a great kid."

In addition to permaculture, Zach saw engaging in urban cultivation as something that could fit into where he was in his stage of life and his priorities at the time. But he also saw the garden as a way to educate the public about the benefits of gardening and to provide symbolic and material resources to community members. The project clearly encompasses all three types of anticipated outcomes, defying exclusive classification into any specific one.

The multiplicity of the growers' ideas for what a cultivation project could deliver also reflected their changing expectations over time, as their projects developed, or when the urban environment in which they were operating changed. For example, even though his gardening project came to be known as a prime example of community-oriented urban gardening project on the West Bank, Bill did not initially start his gardening project with community benefits in mind. He recalled:

And, and to be honest, in the beginning, I really wasn't trying to do outreach. I was just in there on my own and enjoying it, it was my little man cave and I would go in there and grow, and I enjoyed it. . . . But then I started paying attention to what was going on. I volunteered with New Orleans Food and Farm [Network], Parkway Partners, Grow Dat Youth Farm, and going to a lot of the meetings. Some of the things that I became concerned with is that the communities that a lot of organizations work in, they don't really give back to the community. And then that's when I realized I had to change the way I, [how] I approached my garden and, you know, I wanted more involvement. I wanted more, and the biggest thing I learned doing the garden is the need for land management, water management, soil sciences in our youth.

As a Black retiree, Bill's initial interest in gardening was leisure, influenced by his late wife's love of gardening. But he began to see broader potential for his garden as both his experience with gardening and the conditions in the city changed. By more deeply engaging

with urban cultivation, and with increased exposure to other projects' social missions, Bill came to recognize the social impact his
garden could have.

Joel experienced a similar evolution, but in a different direction. His
commercial urban farm, Paradigm Gardens, which he co-founded with
Jimmy, came to be recognized as one of the most successful entrepreneurial cultivation projects in the city by 2015. Despite the economic
success of his latest project, when we originally interviewed Joel in 2014,
he reflected that his earlier experiences in urban cultivation during the
transitional period were primarily focused on social outcomes for the
nearby community:

> Where we lived, just like most places in New Orleans, there's a lot of
> [empty] lots everywhere. And, like certain places in New Orleans, espe-
> cially Central City, especially four or five years ago, there was nowhere
> to really make groceries anywhere, except for like, a corner store, which
> was selling bullshit [quality produce]. So those two things combined,
> yeah, it was something that we thought was necessary, and [what] our
> neighbors saw was necessary. So yeah, we got together and made that
> happen. We did a neighborhood market every Saturday, and we subsi-
> dized that with a twelve-person CSA [Community-Supported Agricul-
> ture]. I own a personal fitness business, mostly [the clients] are middle
> to upper-class people. So we had them on the CSA, which was definitely
> a fair price, but it was enough to subsidize the neighborhood market.
> We charged them a certain amount, and then that [allowed us to of-
> fer] half-reduced prices in our neighborhood market. And some people
> on our actual block would, you know, work, trade, whatever else (for
> the produce).

Joel's previous experiences of working within the community were
likely not evident to those who attended pop-up events or weddings at
Paradigm Gardens. As a young white man that came to New Orleans
shortly before Katrina, Joel's biography would have put him in social
proximity to the YURPs and other post-Katrina transplants. But his case
shows that grower's intentions underlying specific cultivation projects
were not fixed, even if the gardens and the farms appeared to the public
as a constant.

Not all growers had obvious pathways to urban cultivation practices in New Orleans. There were growers that somewhat accidentally stumbled upon urban cultivation practices but came to embrace it after their initial engagement. Among the growers who accidentally found themselves in urban cultivation, some initially arrived to the region as volunteers during the recovery period. Others arrived in the city with plans to do something else and eventually found a new career path in urban cultivation. And there were those whose arrival to New Orleans was itself accidental. For these individuals, their engagement with urban cultivation was not necessarily motivated by their interests or social commitments, but rather a response to unanticipated opportunities or merely out of curiosity. In short, they gave urban cultivation a try—"just to see what happens." Erin came to New Orleans in 2008 after her initial plans to work on a rural farming project fell through. She ended up working with the recovery nonprofit, Lowernine.org, managing their gardens in the Lower Ninth Ward. She already had two years of farming under her belt working in rural agriculture, but a series of events landed her in the city where she learned more about "urban agriculture." She recalled:

> I didn't really know about urban farming. I thought I was gonna be in the country. And then I fell in love with the city, you know? I was there and, I guess you could say I was sucked in a little bit, because of the situation, but it was great. I really loved the Lower Ninth Ward. I loved the people I was working with, rebuilding houses, the neighbors that we were working with. So after a while it wasn't even appealing to go back into the countryside to farm.

Erin helped a long-term gardener from the community develop a farm called Villere Street Farm in the Lower Ninth Ward, placing her on a trajectory of working with several other urban cultivation projects. These serendipitous entry cases, which were not uncommon across the decade, indicate that someone's engagement with urban cultivation was not necessarily premeditated by a concrete set of anticipated outcomes. Anticipated outcomes crystalized as these growers embarked on their projects, confirming urban cultivation as an effective schema guiding practice. In other words, their ideas about what they could do with the cultivation space developed as they dug into the ground, began planting, and saw their social and economic environment shifting around them.

This section has illustrated the range of entry points and goals growers associated with urban cultivation, as well as the multiplicity of these outcomes, even within a single project. But why do growers gravitate toward some goals over others, and what happens when the economic and social conditions that shaped their initial visions for a cultivation project change? In the remainder of the chapter, I look at how a specific type of aspiration for starting a new urban cultivation project dominated the scene for each of the three periods in the decade following the storms.

A Typology of Urban Cultivation Aspirations

To an untrained eye or the casual passerby, most gardens look alike, beyond superficial differences such as which vegetables are growing and whether there are chicken coops or greenhouses on site. There are gardens that post their mission statements by the entry gate or on the fence to inform visitors, yet to many observers, the differences in why these gardens come to exist are not self-evident nor immediately relevant. In the following sections, I present five types of urban cultivation aspirations I identified in post-Katrina New Orleans: *urban cultivation expansion, community rebuilding, alternative food systems, social entrepreneurialism,* and *alternative careers.* This typology distinguishes what the growers had initially hoped to achieve in or through urban cultivation as a form of prefigurative urbanism, based on a few different dimensions: visions for an alternative future, types of outcomes, anticipated beneficiaries of the practice, and how the growers envisioned prefiguring the alternative future through urban cultivation (see Table 2.1).

There were differences in demographic representation among the growers that exhibited each typology of aspiration, as well as differences in the point in time that a certain type of aspiration became more prominent in New Orleans. *Urban cultivation expansion* aspiration was the first to emerge during the recovery period, activated by growers' hopes to encourage more people to grow in the city. During the recovery and transitional periods, two distinct approaches of using urban cultivation

TABLE 2.1. Typology of Urban Cultivation Aspirations

Aspiration Type	Alternative Future Visions	Outcome Types	Anticipated Beneficiaries	Prefiguration through Urban Cultivation
Urban Cultivation Expansion	More gardens and more growers in the city	Social Personal	Current and potential growers	Training more growers Educating the public
Community Rebuilding	Empowered community with access to resources	Social	Long-term residents	Garden as a nexus for social connection Revising the tradition of communal cultivation
Alternative Food Systems	Localized food systems that is accessible to everyone	Social	Society at large	Growing at scale Creating a new food production and distribution system
Social Entrepreneurialism	Alternative economic system outside of racial capitalism Redefining agriculture labor	Social Economic	Marginalized communities	Economic independence (in order to fund the non-profit work) Pay a living wage for agricultural work
Alternative Careers	Making a living as urban growers	Personal Economic	Self	Working as a full-time grower

as a part of the post-disaster recovery emerged: *community rebuilding* focused specifically on the recovery of a particular neighborhood, while *alternative food systems* saw New Orleans or a specific neighborhood as an exemplar of broader social issues or concerns, from environmental sustainability to industrial agricultural systems. The idea of market production slowly began to materialize during the transitional period, but this type of aspiration quickly came to dominate the scene by the redevelopment period. The majority of those who grew commercially were drawn to the appeal of urban cultivation as a pathway to an *alternative career*, but there were few that saw this new economic opportunity as a pathway to create an alternative economic system through *social entrepreneurialism* by blurring the distinction between for- and nonprofit.

The typology of aspiration represents what sociologist Max Weber calls *ideal types*: theoretical concepts that help us distinguish people, actions, places, and institutions in order to make sense of why and how they are different from one another. In reality, each urban cultivation project held multiple aspirations or added more over time. Thus, the illustrative quotes from various growers do not necessarily signify that their project squarely or solely should be categorized into one aspiration type. Instead, the quotes and descriptions of the cultivation projects are presented to exemplify the defining characteristics of each type. What individuals saw as the potential of urban cultivation reflected what they saw as the alternative future that could be prefigured through urban cultivation.

Urban Cultivation Expansion

A small but distinct group of growers found their calling in working on and advocating for the expansion of urban cultivation practices. These were primarily long-term residents and long-term growers who envisioned that their project sites would both demonstrate the potential of growing food in an urban lot and present opportunities to train a new generation of urban growers, whether they were growing for leisure or for a living. I describe this aspiration as *urban cultivation expansion*: the wish to grow more growers. While most growers supported other growers in some direct or indirect ways, from exchanging resources to lending tools, projects reflecting the *urban cultivation expansion* aspiration are distinct for their primary focus on training more people to grow food in the city, for leisure, social outreach, or profit. Their envisioned outcomes ranged from more backyard gardeners across the city to individuals and communities regaining economic and food security through cultivation. These aspirations emerged as early as the recovery period and continued throughout the decade, though the expectations regarding who should be growing and for what purposes varied across the projects and changed over time.

Growers with this type of aspiration often described their projects as "demonstration gardens." Their goal was to set up a garden that would demonstrate to the public the benefits of urban cultivation. Nearby residents or anyone who was curious about gardening could watch and learn, regardless of the scale of gardening they wished to do themselves.

Jeanette, a Black woman, retired teacher, and long-term resident and grower whose flower deliveries to the disaster relief workers opened the Introduction of this book, continued working on the garden that she had already been maintaining since before the 2005 hurricanes. After her return to the city from evacuation, she expanded her work by accessing more spaces, mostly in the Lower Ninth Ward. She described the intent of one of her gardens by stressing how it would serve as an example to others:

> The Garden on Mars was meant to be a demonstration garden to show individuals who are employed how they can garden in a very simple way without being burdened by the maintenance or being concerned about contaminants. It's all raised beds. And the soil is completely covered.

Similarly, Pamela, a Black New Orleans-native and a lifelong grower, described Angel's Trumpet, a garden that she started in 2013, as a place to show the potential of growing food and in the city and cultivate future growers:

> The garden sort of serves as . . . a demonstration site. People are always stopping by, asking how to grow. I'm also working with another nonprofit, Blessed 26. They work to mentor young boys and men in life skills. I hope that a cadre of these boys and men have an opportunity much like the 4H model to get the training here to create their own projects.

The *urban cultivation expansion* aspiration is distinctive for its primary emphasis on promoting urban cultivation to a wider audience and recruiting and training new growers. Macon, who used to answer his phone with a jovial "Macon Fry, the garden guy!," mentored dozens of younger growers in the city since 2005. A white, long-term New Orleans resident, Macon found working with aspiring growers one of the most rewarding aspects of urban gardening. When asked about the motivations for his garden, he focused on exposing others to the positive experience of gardening:

> My first experience in market gardening was having Saturday, tailgate, pick your own stuff here at this garden. And that being such a secluded

neighborhood, it was mostly my pals who I would have given stuff to anyway. But it was just a great way to get people in here, you know? And so once I started at Hollygrove I certainly did not need the money. And so I wanted to get people involved over here. And you know, I forget where we started, but I had a very, very good crew coming here from Hollygrove [Market and Farm volunteers]. We made, we wrote an agreement and I may still have it, and really I wanted, I said, that people, I wanted everybody to have fun, you know? If it ends up not being fun, then we need to revisit the plan.

As a retired teacher, Macon did not view his cultivation work as a source of income or a career, as evidenced by his declaration that he "certainly did not need the money." He saw this as an opportunity to continue teaching younger people and to be a part of the recovery process in a very tangible way. Jeanette and Pamela also collectively mentored dozens of newer growers over the decade, consulting with them on growing technique, financial aspects of starting and maintaining gardens, and providing opportunities for apprenticeship in their own gardens.

The *urban cultivation expansion* aspiration did not just aim to spread horticultural skills but to also model the role that growers and their cultivation sites should play in a community. In this sense, the aesthetics of the garden were important. The garden also needed to be respectful of the surrounding community, because the aim was to make people appreciate and want to support urban cultivation. Pamela emphasized the importance she placed on being an "organized" and "neat" grower, in her approach in working on the Angel's Trumpet Garden, which was located across the street from a popular local diner:

> It's a very public space and right across the street from the Ruby Slipper. It is always my intention with any sort of urban growing to respect the neighborhood and the people that live there. Immediately, I wanted it to be beautiful. I wanted it to be an enhancement for the neighborhood.

Keeping her cultivation site aesthetically pleasing had always been a priority for Pamela, because she saw it as vital to increasing community engagement, which was critical to creating opportunities for the next generation of growers. Here, her visions for the garden to "enhance" the neighborhood should not be conflated with an aim to invite economic development through beautification, as feared by the green gentrification scholars and activists.[5] The purpose of beautification for Pamela was to contribute to the collective efficacy and sense of pride among the residents rather than to increase property values.

Community Rebuilding

The majority of the cultivation projects that started during the recovery and transitional periods were inspired by social outcomes that envisioned gardens and farms as tools for disaster recovery. Unlike the war victory gardens of the twentieth century, however, these projects were not initiated by government policies or funding. *Community rebuilding* aspirations focused specifically on the recovery of a particular neighborhood and in meeting the needs of long-term residents, directly or indirectly, through the gardens, with the aim to rebuild the community's social and ecological resilience. *Community rebuilding* prioritized immediate responses to local challenges, even if the issues had long-term historical roots, such as race-based residential segregation or systemic divestment in Black communities.

It was the community organizers with personal roots in the neighborhood who were most likely to explain their visions of urban cultivation as serving the long-term residents returning to the flood-decimated neighborhoods. Their work in communities that faced severe flooding and displacement focused on various immediate needs of the returning residents, such as housing, food insecurity, and social support, but gardening was rarely the first or the central idea in their community recovery efforts. Nevertheless, these organizers came to see that urban cultivation could provide concrete and symbolic opportunities for rebuilding the community while honoring its history and enhancing its resilience.

These community organizers were not experienced growers, and so it was not their immediate idea to start gardens as part of the recovery efforts. The ideas for the gardens grew out of conversations with community members or were grounded in their knowledge of the history of urban food provisioning in Black and immigrant communities in New Orleans. The gardens would also provide a space to gather and work collaboratively toward rebuilding the neighborhood according to environmentally sustainable practices. Jenga, a New Orleans-born Black woman who grew up in the Lower Ninth Ward, returned to the city after Katrina to start organizing to rebuild her community. She helped establish two gardens in her neighborhood after the residents reminisced about the gardens that had gone fallow years before 2005. She worked closely with long-term residents and linked the project to the rebuilding of the neighborhood:

> Well, the one on Forstall and Chartres was already a garden. It was already a community garden for, I think, about twenty, twenty-five years before Katrina. And after Katrina, when I came back, I had wanted to just, I wanted to do something to help revitalize the neighborhood, and I wanted to put a community garden on my block. . . . I got to know the people who had been involved in [a previous garden on the same site]. Ms. Patsy was one of them, and Ms. Betty who lives down the street, you know. I just kind of learned more about the history of this garden, and started organizing people.

Despite growing up in the neighborhood, she was made aware of the history of the old garden site through her conversations with neighborhood elders. As noted in Chapter One, Jenga's generation did not see a robust urban cultivation practice in New Orleans as they came of age in the 1980s and the 1990s. In contrast, Pamela, who is a generation older, had firsthand experiences with and consistent memories of the practice throughout her lifetime in New Orleans. While Jenga had always intended these gardens to serve the interests of the community, preparing the garden itself became an opportunity for community organizing. Volunteers did door-to-door outreach and held events to gather residents, and the neighbors came out to show support.

Because food production was not necessarily the primary focus of the *community rebuilding* aspiration, gardens and farms were also expected to provide community members with spaces for gathering and socializing. The food being grown and harvested was intended for communal distribution on a smaller scale, not necessarily to feed a large population or to provide economic opportunities for the growers. The director of the Center for Sustainable Engagement and Development in the Lower

Ninth Ward described the plan for an orchard garden that was in the early phase of implementation in 2014:

> They're community-based gardens and farms. For example, on Dauphine Street, we don't have any fence around it. So it's designed to be accessible to the community. Anyone in the community will come in, to be able to take advantage of the walk-in garden. . . . They can, you know, ponder their thoughts, in a kind of nice environment. We don't have—there are no restrictions on that basis. Then, once we have harvest of the orchard, you know, which we will give away to the community—we haven't figured out how we are going to do that yet, it just depends on how big the harvest is. But, you know, it's all designed to be a benefit to the community, an example of what can be done in the community.

The site this director was describing would house a combination of fruit orchards and butterfly gardens. Along with a sustainability education center that the organization was planning down the street from the garden, the goal for the garden was to invite community members into the space as volunteers and as users. The director, who is Black and has familial roots in the neighborhood, also noted that the gardens could contribute to the sustainable redevelopment of the community by converting blighted properties into maintained green spaces.

While *community rebuilding* was no longer a prominent aspiration type by the redevelopment period (2012–15), community organizers would still identify a need for gardens for local residents in the areas that continued to face economic and social challenges. For example, a Black former police officer had been trying to start a garden for the community members of his Central City neighborhood on the lot he had leased from NORA a year earlier, in 2014. When asked about the kind of inquiries he received about the garden, he responded:

> Oh, yes. Oh, I have people in line. I have a couple of people just around the corner, you know. They're ready. I mean, they have knowledge. I have one lady, and it would help her so much, because she's part of, uh, should I say, the drug community, but she has knowledge in plants and growing, and she's begging. She said, "That would give me something to do. It'll keep my mind off of just running." I'm like, "Boy, I sure wish I had it, too."

His vision for community recovery did not focus on the 2005 storms, but on the long-term economic and social disparities that faced the community where he was born and raised. He hoped that the garden would provide a positive outlet for community members while also creating a *safe* space that contrasted with houses in the immediate vicinity, where criminal activity related to drug trade was prevalent. As we sat in his pickup truck to conduct our interview, in front of the future garden site, he pointed to a security camera installed on the lamp post in front of the property and recalled the most recent shooting incident in the neighborhood, which took place in a house adjacent to the site. He felt that he was creating something that many people, especially seniors, in the community really wanted but lacked resources and capacity to develop on their own.

The *community rebuilding* framework, for some growers, meant honoring the long history of Black urban cultivation in New Orleans. To those who remembered the ubiquity of backyard and porch gardens and the communal sharing of food among families and neighbors, the practice embodied one of the many ways that the community retained resilience in the face of discrimination and segregation, social disasters that have plagued the city and the nation for centuries. The post-disaster city seemed a most apt situation for revising the tradition by reinstating the practice. Pamela, whose project primarily focused on *urban cultivation expansion*, was one of the few that articulated this as a grower during our initial interview. She explained why she decided to name one of her projects Angel's Trumpet:

> I named it Angel's Trumpet because I love the old-fashioned Southern coastal flower with its beautiful downturned blossoms that you see a lot around here. My hope is that people here understand that gardening on this level is a revival; it's nothing new. I grew up with family members with teeny gardens in their Uptown New Orleans backyards. We had amazing stuff growing because a lot of people had rural and immigrant connections. In every community, from the Italians to the Irish to Black to everybody. I grew up with gardens and fruit trees and that was just part of it. That was part of the natural landscape here. So, I think more about revival.

Viewing urban cultivation as a *revival* rather than a new idea distinguishes *community rebuilding* from the *alternative food systems* aspiration. As we will see in the next section, the *alternative food systems* aspiration relied on the assumption that there was not a previous practice of growing food at scale in the city, especially in Black communities where many of the growers planned to start their new cultivation projects. Notably, long-term and multi-generation New Orleanians were not necessarily aware of this history, due to the long-term suppression and decline of these practices prior to the storms. Many new growers learned of the history in the process of working on their projects, through their conversations with community elders or when working with mentors like Pamela.

Alternative Food Systems

Emerging during the recovery period, the *alternative food systems* aspiration envisioned using urban cultivation for addressing growers' concerns about broader environmental and food systems issues. This aspiration aligned closely with the alternative food movement that grew out of the popularity of bestselling books such as Michael Pollan's *The Omnivore's Dilemma* or Barbara Kingsolver's *Animal, Vegetable, Miracle*. The central concern of this movement was the environmental and health ramifications of the agricultural-industrial complex, and it emphasized individual consumer actions to eat fresh, local, and seasonal as one of the most immediate solutions to these issues. Around 2005, food justice scholars began critiquing this movement, which they often described as alternative food networks (AFNs), for its lack of reflexive awareness of the privilege inherent among many of its proponents, and for reinforcing the neoliberal framework of individual accountability within the marketplace, rather than calling for systemic structural changes in the agricultural industries and food supply chains.[6] But in 2010, when the *alternative food systems* aspiration was motivating the establishment of new urban cultivation projects in post-Katrina New Orleans, such critique still remained in the academic field or among grassroots political activists with whom most of the growers rarely engaged.

Growers with the *alternative food systems* aspiration viewed the city, or a particular neighborhood within the city, as a microcosm of the broader American society, with a particular focus on food insecurity and environmental sustainability. At the core of this aspiration type was a desire to bring to New Orleans the latest research on and innovations in sustainability practices developed elsewhere. These were *new* issues to many of these growers, in contrast to the *community rebuilding* aspiration's focus on reviving the historical local food provisioning practice. The ideas and the knowledge behind these visions were acquired through formal education and training, or general interests in climate change and food systems' impact on environment and health. The growers with this type of aspiration saw the post-disaster city as an exemplar of universal social issues and sought opportunities to implement alternative futures in which New Orleans would catch up with or even precede the rest of the world in addressing food-related and environmental social issues.

While most of the post-Katrina transplants starting cultivation projects during the recovery and transitional periods expressed the *alternative food systems* aspiration, this was not the only group that held this aspiration. Some younger New Orleanians, both Black and white, also referenced ideas about sustainability that were grounded in theories like permaculture. Jordan, a white New Orleans-native, developed an interest in environmental sustainability and permaculture during college, where he "did a lot of environmental organizing." After college, he participated in Worldwide Opportunities on Organic Farms (WWOOF) on a couple of rural farms in Louisiana and "had a blast" learning about organic farming. At one of the farms, he met another gardener who helped spur his strong interest in permaculture. He recalled:

> We kind of shared this vision. I'd heard of permaculture, I guess. I don't know, at some point in college. And [it] sounded like the greatest idea I'd ever heard of, because it seemed like a solution to a lot of environmental issues we have. And a way to provide for human needs while also restoring ecosystems. Two things that I really wanted to be able to do. And he's really into it as well. We kind of had a dream of starting this permaculture, naturalism education center out there.

After the project they had launched in Jackson, Louisiana, was discontinued due to land access difficulties, Jordan returned to New Orleans to work on landscaping and nursery, and eventually started a project that focused on teaching do-it-yourself (DIY) gardening and permaculture education, reflecting his *urban cultivation expansion* aspiration. His interests and knowledge were developed through his exposure to the topics during college, in addition to hands-on learning in rural agricultural settings.

The *alternative food systems* aspiration considered urban cultivation as a tool to address a variety of social issues. For example, growers with this aspiration saw growing food in the city as one direct and concrete way to address the "food desert"[7] problem, a concept describing the concentration of food insecurity in communities of color and low-income neighborhoods that was gaining public awareness around 2005. By the transitional period, it became clear to many that grocery store access was unevenly distributed across the city. Areas that were recovering quickly saw not only reliability of business hours and stock at existing grocery stores, but also benefitted from the opening of new grocery stores such as Robért on Carrollton in Uptown and Rouses in the Central Business District. In contrast, the areas struggling to rebuild had no grocery stores reopening, nor opening anew. Lower Ninth Ward residents were forced to travel across the industrial canal bridge toward downtown or to the Walmart Supercenter in the neighboring St. Bernard Parish to "make groceries," a local vernacular term for grocery shopping, with no full-scale grocery stores operating in the neighborhood.

Concerns about broader food systems or environmental issues often drove the *alternative food systems* aspiration, and specific projects were conceived of to attain the larger goals of reconstituting the existing economic and social systems. Yet none of these growers aspired to feed the entire city through urban cultivation. After all, most projects did not expand beyond a few urban lots. These growers viewed the idea of growing food in the city as both symbolic and concrete, a small-scale challenge to industrial agriculture. Similarly, the association held between urban cultivation and environmental sustainability was optimistically prefigurative of broader social changes by directly engaging in public education and local ecological mitigation efforts. Gardens were expected to be a part of a sustainable urban landscape by providing multitudes of ecological benefits, from rainwater runoff prevention to soil remediation and blight overgrowth management.

Another sentiment associated with the *alternative food systems* aspiration was the belief that the public should be encouraged to think about these social issues in a context beyond New Orleans or their specific neighborhood. This line of thinking presumed that the public lacked knowledge beyond their immediate experience, or that their inaction toward these issues resulted from ignorance. Colleen, a white post-Katrina transplant, described her motivation to engage in urban cultivation:

> I really want to have people understand how to grow food. My passion is not to be a farmer. My passion is to teach this food thing that we're all obsessed with. We're all obsessed with food, but none of us know how it grows. Very few of us have actually ever done the work. . . . It's like when a bunch of people who are never at the table are actual farmers. Imagine if we had all the immigrant laborers at the table when we really talked about how we wanted to change our food system, you know? So, for me, I'm trying to reach out to educated people to say, let's really talk about access.

The intended audience for her work was the "educated people" that she felt needed to be made aware of problems related to food production or food access, but the goal was to raise awareness of the issue in hopes that those with economic means would then take necessary action as consumers to change the system.

Growers who held the *alternative food systems* aspiration tended to envision larger societal changes, even if they were to be enacted at a local level, but often they had not identified concrete pathways to achieving these goals beyond their own projects. These grand visions reflected their own values and interests, as these growers dreamed big and attempted to prefigure an alternative social system, from food production and distribution to urban land use and non-traditional education systems. This contrasted sharply with the concerns expressed by those who were motivated by the *community rebuilding* aspiration, whose interests in urban cultivation derived from their wish to empower a particular neighborhood, or *urban cultivation expansion* aspiration that focused more specifically on more people gardening for various reasons. For growers motivated by the *alternative food systems* aspiration, concerns for the broader issues came first, before they even identified suitable sites in post-disaster New Orleans for implementing their ideas.

The common driving force behind the *alternative food systems* aspiration was the notion that the city or specific communities were in need of "help" and that gardens were going to be a part of the solution. These sentiments echo what came to be commonly described as "white savior complex," a trope in which a single white figure or white-led organization comes to the rescue of those who are perceived to be helpless on their own. Such a disposition fits seamlessly with the post-Katrina influx of well-intentioned saviors into New Orleans. From celebrities like Brad Pitt to national nonprofit organizations like Teach for America, volunteers arrived in droves to help the city that seemed abandoned by the federal and state governments. Yet programming with this orientation prioritized volunteers' sense of accomplishment rather than the actual needs of local communities.[8] Community organizer and author Jordan Flaherty offers an examination of the negative impacts of these (white) saviors when they arrive in these communities: the problem and the solution are defined through the eyes of the outsiders, and their capacity to access and mobilize resources make these newcomers more powerful than long-standing grassroots efforts that become overshadowed and even compromised.[9] C. W. Cannon, a New Orleans-native and author, articulated this tendency as follows:

> Until they acknowledge the New Orleans that existed before they got here, they will simply be colonialists, imposing whatever their uninformed and youthful imagination wants on what they falsely perceive to be a blank slate. [New Orleans] natives have opinions, too, y'all, and have been engaged in defining their own city for generations. The national press shows, as often in the past, that it's not really interested in New Orleanian self-analysis. Apparently we're not qualified.[10]

Some growers with *alternative food systems* aspirations exhibited characteristics of the savior complex: presuming to know the community's needs, the view of urban cultivation as *new* and unprecedented, and inflated expectations about the scale of the imagined social changes that may be achieved through the gardens and the farms. For example, a community's designation as a "food desert" justified the placement of an urban cultivation project without a clear indication of whether or not such a project was explicitly or implicitly desired by the community as a solution to the problem. A white founder of a nonprofit organization that focused on promoting hydroponic and aquaponic growing systems described why she decided to headquarter the organization in New Orleans:

> When we were starting the organization, we looked at different locations around the country and sort of overlaid a bunch of different statistics— that was poverty, food deserts, differential between low and high income in different areas, and a number of other things [such as] obesity rates. And [we] came up with five or six different possibilities for a headquarters for [the organization]. . . . The day I got here, there was a big press conference at Hollygrove [Market and Farm], and the Rethink school kids were there. And they were talking all about needing healthy, fresh food in schools, and kids were tired of eating lousy lunches, and they were tired, and they were overweight. They wanted healthy, fresh food. And I thought, okay, I get it, you know, this is the place. So that's how we got to New Orleans.

The idea for the organization was there first, and New Orleans was identified as an appropriate site for implementing it. Despite the references to local advocacy for an overhaul of the school lunch program, the primary operational goals of her nonprofit were to provide legal and policy support for urban growers, not to work directly with the school system or local activists on food justice issues. Abstract and hopeful descriptions of the cultivation projects as "the solution" reflected external *diagnostic framing* ("What's the problem?") and *prognostic framing* ("What's the solution?") of the complex social and economic challenges these communities were facing.[11]

The *alternative food systems* aspiration's grand and broad visions for change were another manifestation of the savior trait. Nat Turner, the founder of Our School at Blair Grocery in the Lower Ninth Ward, describes one of the most important aspects of working with young people in the documentary film *Reversing the Mississippi*:

> First thing you gotta do is to agree that you're gonna love 'em like they're your own. Don't come in on a missionary mission. Don't come in to save them. Don't come in to fix what's wrong with them. Learn to love who they are, as they are, with whatever faults, I help 'em.[12]

Turner, as most everyone referred to him, could have drawn these perspectives from his previous work as a high school teacher. Yet when it came to the potentials of the urban cultivation project, he described a few grand visions in the documentary, such as, "It's an earnest attempt to save the world," and "On several acres we can provide food security for the entire Lower Ninth Ward." Turner, who is Black, initially came to New Orleans with groups of New York City high school students to assist in the recovery efforts. When he decided to stay in the city, he envisioned an alternative school for the young people who were being failed by the existing public education system, eventually making urban cultivation a central theme for Our School at Blair Grocery. I also observed Turner share similar and somewhat hyperbolic and grand optimistic visions in his public speaking engagements or interactions with visitors that were often quite effective in getting their attention and support, though these statements were often coupled with a sobering recognition of the challenges in achieving those goals.

To be clear, growers with the *alternative food systems* aspiration cared deeply about the well-being of the long-term residents and the communities experiencing slow rates of recovery, and envisioning the impossible as a starting point of prefigurative urbanism. But these growers' visions for change were informed by their knowledge of the solutions developed elsewhere to be imported to and implemented in New Orleans. Specialized and academic concepts such as permaculture, green infrastructure, or food justice often accompanied these large-scale aspirations, and with them often came a level of confidence among these growers that they *knew* in advance what the solutions would be. In many cases, the projects and the growers evolved over time to reflect on this type of aspiration, as their desire to contribute to the betterment of the community or the society remained constant.

Social Entrepreneurialism

During the transitional period (2009–11), many urban cultivation projects began engaging in market activities. There were some very limited commercial activities prior to this period, in the form of informal CSAs or farm stands, though this was very experimental and sporadic. The

CSA model, a subscription-based system, typically requires advance payment from customers at the beginning of the growing season to farmers, with an expectation that the farmers will provide produce throughout the growing season in return. The opening of two new alternative food markets and renewed support from NGOs for commercial urban cultivation around 2010–2011 created a new set of opportunities for the growers to envision a robust, locally grown produce market. The first group of gardens and farms to engage in commercial production were the nonprofit projects that began during the recovery period with the *urban cultivation expansion* or *community rebuilding* aspirations. They used the revenue to supplement their financial sources, thus easing their dependency on grants or personal savings. By 2015, several cultivation projects were set up as for-profit enterprises to engage in commercial growing with social missions. I categorize this type of aspiration as *social entrepreneurialism.* A strict definition of the term *social entrepreneurialism* has been elusive, but the term generally refers to businesses and business people that place equal value on their social impact and financial bottom line.[13] In this study, both for- and nonprofit cultivation projects were associated with this type of aspiration. The term *social entrepreneur* was rarely used by the growers themselves, partly due to the novelty of the concept in the city at the time,[14] but this type of aspiration embodied the concept's blending of activism (social outcomes) and entrepreneurialism (economic outcomes).

The *social entrepreneurialism* aspiration subsumed a variety of approaches to social change. Profit from market sales could be used as a funding source to support primarily nonprofit cultivation projects. David, who is white, initially came to New Orleans to volunteer as part of the recovery effort. After deciding to stay and start his own organization to continue recovery work, he initially envisioned focusing on housing, but due to the bureaucratic challenges in getting that project off the ground, he turned to work on food insecurity. Based on his previous experience in rural farming, he considered urban cultivation as a way to address food insecurity in the neighborhood and began growing on his first vacant lot in 2010. As the project expanded in its scope and size, his nonprofit organization, Capstone, began selling produce and honey. When asked about the organization's mission during a 2014 interview, he described it as follows:

We have a very broad mission. Some of them stem personally by me. The most visible thing we do is to grow food and give it away to people in need. . . . We helped with the farmers market [in the neighborhood] recently, trying to assist Burnell to get the first grocery store in the lower Ninth Ward open. So we donated 100% of what we took [in] over there to him. And another place [that Capstone engages with], what we do is we also take food over to Thompson Center in the city. They're a daytime homeless shelter. And then we do the honeybees and raise them, sell the honey to help pay for the food to give away.

The organization's model of selling some of the produce to offset the cost of donating or selling produce at discount to nearby residents was not unique. It was a common practice of nonprofit cultivation projects. David's reference to the homeless shelter and to Burnell Colton, who was about to open the first grocery store in the Lower Ninth Ward since the storms, also shows that the funds were not simply to support the organization's own operational costs but were used to support other recovery efforts in the neighborhood.

Nonprofit urban cultivation projects' forays into market activities were also motivated by their efforts to secure financial stability and independence. Our School at Blair Grocery was one of the first nonprofit cultivation projects to carve out a customer base within the city's fine-dining scene. The organization had been expanding its cultivation spaces around its original location through a combination of lease and purchase, and Rob, the co-director, realized, "We could use that to generate money as opposed to getting grants and donations and things." Youth education in urban cultivation also included entrepreneurial training, as was the case at both Our School at Blaire Grocery and Grow Dat Youth Farm. Grow Dat Youth Farm used urban cultivation to provide paid leadership training to local high school students from a variety of backgrounds. Young participants in these educational programs were taught not just how to grow but also the economics behind agribusiness, the food distribution chain, and social justice movements concerning food and environment. Thus, the organization's direct engagement with the market was a demonstrative learning opportunity that simultaneously created an additional income stream for the organization while meeting its youth empowerment mission.

There were growers that decided to opt out of nonprofit status, primarily for the operational and financial autonomy associated with for-profit status. The growers who expressed this type of *social entrepreneurial* aspiration during the interviews were mostly younger, post-Katrina transplants with strong left-leaning, or even radical, political views. These growers often used market approaches to support alternative economic forms that acted as a critique of capitalism and the mainstream economic market system. Jamal started Supporting Urban Agriculture (SUA) in the Lower Ninth Ward in 2011, and described the hybrid nature of his for-profit cultivation project:

> We are going to be going through restructuring, but the reason why the LLC was created as an offshoot of something else (was because) it was just me. The idea was to try and get it to be a collective, cooperative of some sort. Its mission is to create a sustainable economic model for selling, growing, buying practices. But we do a lot of nonprofit work, by definition. Not for credit, but by nature of, I'm not saying we're great, but it's hard for us not to think like capitalists, and we're not communists either, but I don't know . . .

Jamal, who is Lebanese and Syrian American, was primarily motivated by his concerns about environmental sustainability and structural racism. But he opted to establish his project as a for-profit enterprise rather than a nonprofit organization, partly to have better control over his practice. "Sustainability" in his statement above refers to the economic viability of urban cultivation, even though he was equally invested in ecological sustainability, and he found that being a limited liability corporation (LLC) gave him more autonomy and flexibility to pursue the kind of model that he was trying to accomplish: a cooperative or collective economic system that is distinct from both the traditional nonprofit organizations and conventional for-profit businesses.

Projects with the *social entrepreneurial* aspiration also aimed to create economic opportunity through jobs for people in the community. Unlike the *alternative career* aspiration, the aim was not necessarily for the business owners to establish their careers but to empower long-term residents. It is also distinct from the *community rebuilding, urban cultivation expansion,* or the *alternative food systems* aspirations for its explicit focus on economic outcomes, though all aspiration types were motivated by some social outcomes. When asked about the primary goals of Good Food Community Farm, a cultivation project that began around 2012 in the Tremé neighborhood, Cory, one of the co-founders responded:

> I mean, employment, you know? Our big focus is on growing the business so that we can do more (for the community). . . . We work with Covenant House. We work with Food and Farm Network and Tremé Community Garden. But we're totally funded. We're a small business with five employees, and we're funded through sales.

Cory is white and a post-Katrina transplant, but part-time growers employed by Good Food Community Farm were Black long-term residents from the neighborhood. While employing a few people to work on their commercial cultivation sites might have a limited economic impact, it still held the potential to yield a concrete and immediate outcome. The organization's use of paid labor was also in contrast with the reliance on volunteer labor among many nonprofit urban cultivation projects, which some growers saw as exploitative or ineffective, as I illustrate further in Chapter Four.

Overall, the *social entrepreneurialism* aspiration was more closely aligned with the *community rebuilding* and *alternative food systems* aspirations than with the *alternative career* aspiration. Social entrepreneurial growers' decision to conceptualize their projects as for-profit enterprises was due to their perception of the limitation of the nonprofit structure. These growers were making these decisions on the heels of the 2009 financial crisis and the 2010 BP oil spill, and the changing priorities among many national foundations reminded the growers of the risks of relying on grants, leading them to view market cultivation as a source of funding that offered stability and independence. As one grower noted, "As long as we don't have any kind of permanence in grant or governmental support or institutional support, I think any kind of nonprofit doing this thing is gonna be more impermanent than somebody trying to make a business out of it." Many of these growers were once part of nonprofit organizations and saw running a small business that they could closely manage with economic independence as a more promising path for making concrete changes in the community, if at a much smaller scale.

Alternative Careers

By 2015, newer for-profit projects became the most prolific type of urban cultivation in New Orleans, led predominantly by younger post-Katrina transplants with a college education. The growers seeking to establish themselves as entrepreneurs found themselves stifled or dissatisfied in their previous occupations and saw urban cultivation as an opportunity to change course and pursue their passion. Caroline, who is white, initially moved to the city in 2007 following her work in the film industry. She had already been community gardening for many years, and as she was beginning to rethink her career trajectory, she decided she was ready to "graduate to the farm" and launched Grow Me Somethin' as a business. Coincidentally, an acquaintance of hers knew of an organization that happened to be looking for a full-time grower for their existing gardening site, which allowed her to establish quickly in a space that had already been set up for growing. When asked about why she decided to pursue cultivation full-time, she responded:

> Well, it was driven by a couple of things. I loved my career in film and television, but it is very unreliable . . . and the gardening during the last five years has (been) sort of like the place I'd go when I wasn't working. So I've been looking for my next career, you know, consciously looking for my next career . . . and right at the time the [organization from which she was subleasing the land] finally gave me the okay and handed over the key to the property. My husband had a very good job that allowed me to really last three months, focusing my energy entirely for this [without having to work another job]. Which is, you know, timing was everything, in so many ways.

Caroline was slightly older than most of the newly established commercial growers at the time, but her sentiments of wanting to try to make it as a full-time urban grower as an alternative to a previous career path exemplifies the commonly cited appeal of urban cultivation as a profession. Growers like her often had some previous growing experience, but this was their first time making urban cultivation their primary income source. Feeling unsatisfied with their previous work experiences, especially lack of control over one's own schedule, they sought a new challenge, one that would produce tangible outcomes in a relatively short amount of time. But as we see in more detail in the next chapter, Caroline was in the city as the urban cultivation scene was rapidly growing, and there were pre-established markets for her to sell her produce. Her aspiration to become an entrepreneurial grower was activated by the external circumstances that allowed her to imagine the possibility, from land access to commercial transactions.

The *alternative career* aspiration was not just about changing careers. It was also about altering growers' relationship to work. This involved restructuring the way they lived by turning a job into a commitment to a broader lifestyle. Younger full-time growers attributed urban cultivation's appeal to the *urban agriculture imaginary* of bucolic, utopian, and *real* work, without compromising on the convenience of urban life.[15] Nicole, a white post-Katrina transplant, who operated a flower farm in New Orleans East, described her motivations for starting her project:

> The reason I started it, it's really because I wanted to garden, which seems like kind of a selfish reason to go to all this trouble, but just really this is what I want to do. And I wanted to grow my food and do it with my friends and neighbors. I always secretly wanted to live on a commune and have everybody, like, be farmers, but I also want to live in New Orleans. So this is the closest thing I can get to that.

Nicole sought a career shift that allowed her to do what she really wanted to do. For her, farming was not just a way to produce an income, but also a lifestyle. Her description of the work of farming was notably detached from an acknowledgment of its racial and exploitative history in the American South. Rather, she understood farming as a pathway to a richer life, even with the inherent financial risks. As Caroline's passing comment about her reliance on her husband's income indicates, not everyone could afford a chance to "give urban cultivation a try." It is not surprising, then, that the two demographic groups that were most likely to pursue an alternative career through urban cultivation were younger adults and retirees.

These growers were almost all post-Katrina transplants, and they seemed to have come from a relatively non-traditional work history prior to entering urban cultivation. Among those who expressed *alternative career* aspirations, previous occupations ranged from limousine driver to personal trainer to carpenter. Except for the retired teachers, most of them had a history of switching jobs rather than staying in one industry for a long period, or working in an unpredictable gig-like set-up. Eric, who's white, began his engagement with urban cultivation while he and his friend were living in an RV parked conspicuously at the back of the community garden space of Hollygrove Market and Farm. By the time I interviewed him for this book in 2015, he was growing commercially on the West Bank to provide produce to Café Hope, a non-profit culinary arts training program affiliated with the Catholic Charities. He did not come to New Orleans expecting to work on urban cultivation, but he quickly found a new opportunity in working within the emerging scene by finding opportunities while learning on the job:

> I've always been a working-type person, if you will, my whole life. And I've never, like—I'm the kind of person that you know I could be making a hundred dollars an hour yesterday and [still] quit my job, you know? I've owned my own companies, I've done a lot of variety of things. So, I don't know. I'm just a person who, technically, what I tell people is I'm on a working vacation in New Orleans, and I've been doing it for five years. But you know, I've filed my taxes for the last year, and I've made 15,000 dollars. But I also didn't pay rent, I don't pay electricity, I don't pay heat you know? I eat like a king, mostly, ever since I've arrived in this town, you know. So it's really hard for me to classify myself.

After moving his RV to another location, he continued to live out of it and could dine at the restaurant for which he was providing produce. The low cost of living and the ability to eat well seemed like a good deal to him, though his reference to being on a "working vacation" indicates he did not see this as his permanent career trajectory. Growers often candidly shared their uncertainties about the long-term prospects of their project, but underlying that observation was some sense of confidence that there would be other opportunities for them if this path did not work out.

Conclusion

The growers in post-Katrina New Orleans came to urban cultivation with a variety of aspirations, some that the growers did not even realize until after they'd begun growing. I opened this chapter with a description of a range of benefits that the public tends to associate with the gardens, and I speculated whether these metaphorical connections between the gardens and recovery were what mobilized urban cultivation in the city as it underwent major transitions. In reality, the growers were not naïve dreamers who saw urban cultivation as a panacea. Their aspirations for urban cultivation envisioned tangible and direct outcomes, not abstract, long-term ideas of what the future could be. Something about urban cultivation ignited inspiration and guided their visions for what they hoped to achieve through gardening and farming in the post-disaster city. These initial aspirations may have guided their foray into urban cultivation, but their ideas about what is possible through the practice continued to change, as the next three chapters will illustrate. The growers approached their projects as ends-effacing but without a fixed set of preconceived outcomes. Throughout the process, the concrete, practical aspects of urban cultivation guided their decisions, allowing them to add, modify, and reprioritize their aspirations and actions as they went.

This chapter demonstrates that urban cultivation as a schema-guiding frame has a wide range of possibility and can take various forms, allowing for its alignment with various and sometimes competing values, from radical activism or leisure to entrepreneurialism. These aspirations are not inherent in urban cultivation; rather, urban cultivation served as a resonant "cultural repertoire" through which to make sense of unsettled social and personal conditions. Applying Ann Swidler's theorization of "cultural repertoire," we can say that the growers used their varying, emerging ideas about urban cultivation to see what else could be done to bring about immediate changes through growing plants and animals and cultivating soil. Growers who knew what outcomes they wanted to deliver, be it social gathering space or food security, found it easier to see how urban cultivation fit into those desired outcomes. But even for those who were driven primarily by their interests in gardening or farming itself, the practice led them to recognize the kinds of social changes that it could bring about. This was one of the reasons why post-Katrina urban cultivation emerged as a form of prefigurative urbanism, rather than part of conventional social movement. The growers all saw it as a way of making changes, but for so many different reasons and ways.

The urban growers' aspirations were not merely expressions of their perennial inner desires but reflected their personal framing of the issue they wished to address through urban cultivation at a particular time, including personal life-stage contemplation.[16] Gardening appeals to different types of growers, and this appeal ranges with their economic and social circumstances, just as it depends on the outcomes they seek, including personal ones. This explains how urban cultivation seamlessly adopted a different set of aspirations as the city went through a transformation from recovery to redevelopment over the decade. It also explains why the practice remained a collection of individualized actions of prefigurative urbanism, and did not coalesce into a collective activism around a unified cause. The next chapter will illustrate the process of turning these aspirations into reality. As we shall see, hope and grit were not enough to manifest urban cultivation; it took the right location, the right time, money, experience, and even some luck.

3

Breaking Ground

Manifesting the Alternative in the Existing Social System

Caroline, who described her entry into urban cultivation as, in part, a product of perfect "timing" in the previous chapter, was one of the lucky growers who found themselves in the right place at the right time when she decided to start a new cultivation project. When she told me how she got a start on her commercial farm, Grow Me Somethin', from breaking ground to making sales in a matter of months, she emphasized her good fortune in finding a site that already had a garden infrastructure. In particular, that the garden had access to water through a meter associated with a nonprofit organization that had a building on site was especially important:

> Because it's not separately metered, they haven't had the thought of splitting the meter yet. And I also inherited a fair amount of drip irrigation system [that] was already in place, the rows were built. I walked into a farm that hadn't been worked for six or eight months, but had all the parts ready there. So I walked into a very, very sweet situation. . . . I had to augment to buy all the seeds and whatever tools that I've acquired since then. But it's cut my start-up time and my start-up cost dramatically, because within two months I was selling. I got out of the gate in a way that was unusual for anybody in my situation.

Hers was an ideal situation, with access to a space that had already been a garden and that required very little financial and physical investment to be ready for growing. Caroline's appreciation of the benefits of this site were informed by her prior attempts to acquire a site. About two years prior to our original interview in 2014, she had attempted to acquire a property in Pontchatoula, LA, about an hour north of New Orleans, but could not complete the purchase because of mortgage complications with the bank. She had also bid unsuccessfully on multiple lots held by the city's housing authority prior to locating this space through an acquaintance who connected her to the nonprofit organization that was seeking someone to tend the lot.

In my conversations with urban growers, it was much more common to hear stories of failed attempts to launch cultivation projects than successful ones. Shannon, who is white, had lived in New Orleans on and off throughout her adult life, but had returned to the city in 2011 with plans to pursue a master's degree. She soon changed her mind: she wanted to be an urban farmer. Shannon had never farmed before, though she comes from "a long line of farmers" in the Midwest. After interning with Macon at Hollygrove Market and Farm and Gathering Tree for a year, she and her growing partner began looking for their own space in which to grow. The search would not be easy. Shannon and her growing partner first tried to access a site they located through "Farm This Now!," a website operated by New Orleans Food and Farm Network (NOFFN) that listed vacant lots in New Orleans that could be used for urban cultivation, mostly by lease. They found a privately owned lot in the Freret neighborhood and began the process of gaining access just around the time the area was starting to show signs of redevelopment after the opening of an upscale cocktail bar. The land owner dragged out the process, but Shannon and her growing partner began clearing the lot in good faith with the trust that the lease would be formalized eventually. Unfortunately, the lease was never signed; as soon as the land was cleared, the owner reclaimed it, and Shannon and her growing partner lost access to the space.

After that experience, Shannon and her growing partner attempted to buy a lot in the Hollygrove neighborhood. This process turned out to be even more complex, reflecting the vulnerability of landownership in the city, especially in predominantly Black, working-class neighborhoods of New Orleans:

> So the person bought it on a tax sale and had been maintaining the property for, like, five or seven years or something like that. So he was prepared to sell it to us. But when we went through the process, [it turned out that] he actually hadn't been paying the taxes on it, and he actually didn't own the land. So he couldn't sell it to us because he didn't actually own it, even

though he had been maintaining it for all of those years and was told that he did. But when we dug down deep, he did not. So we didn't buy the property because we were not prepared to deal with all of that.

Amid this failed purchase, Shannon and her growing partner started a project in collaboration with a new development project in Mid-City led by a community development organization. In 2014, they developed a gardening space in partnership with various local nonprofit organizations working on issues ranging from education, employment, and affordable housing to food access, which included a high-end grocery store. The aim of this garden was to engage community members as part of a larger project that focused on community health. After the previous failed attempts, Shannon and her growing partner were drawn not only to the mission of this project, but also to the legitimacy and stability of the land access.

∽

What does it take to start a new urban garden or a farm? This chapter chronicles the challenges that aspiring growers faced in trying to accomplish the fundamental first act of urban cultivation: gaining access to a growing space. What new growers in New Orleans quickly learned was that not all vacant lots were *accessible*, and the ones that *were* accessible did not necessarily meet their needs. Finding accessible lots that matched their needs was just as complicated as any real estate search, but with added complications that were specific to urban cultivation. Location, cost, and conditions of the lot varied depending on what the growers were hoping to do with it, and even with a perfect match, the actual processes of establishing access and preparing the lot for production took additional time, money, and energy. Even if growers could access a site, preparing it for cultivation could involve removing debris, establishing water access, and remediating and building soil, all of which took months if the growers wanted to do it naturally or on a budget. Beyond finding and securing access to a space, the growers faced persistent challenges that derailed or delayed their plans. They had to navigate a chaotic and inefficient bureaucracy of government offices that were often unsure about what to do with a form of land use the growers were envisioning.

Throughout the decade, starting a new urban cultivation project proved to be challenging, but for different reasons during each period, reflecting the rapidly transforming economic and social contexts of the post-disaster city. The growers' focus remained on the direct action of growing, so they prioritized fixing what needed to be fixed along the way in order to start digging and planting. They were prefiguring an alternative land use, and the city's legal and economic systems did not yet recognize such practice. They also tried to retain their autonomy through selective collaborations with external interests, rather than merely jumping at any and all opportunities presented to them, especially when others began to see the value of urban cultivation during the latter half of the decade. Trying to start a new cultivation project became a first and significant step for growers to demonstrate how to manifest practices that do not fit into existing forms of urbanism. As they faced and overcame hurdles in turning their initial visions into reality, the growers articulated "willful hope" through their actions while maintaining a level of pragmatism.[1]

This chapter illustrates the range of structural hurdles the actors who implemented prefigurative urbanism encountered; in particular, I focus on how the opportunities and constraints for these practices changed over time and how the actors responded to these external changes. The growers were not discouraged by the initial hurdles, even if they felt frustrated or lost at times. They seemed to accept that such challenges came with trying to do something new and different, and so they soldiered on. Despite the similarities of their challenges, however, they continued to approach their challenges individually, seeking solutions that would sufficiently remove obstacles for their projects to carry on, but without mobilizing around their shared grievances.

Finding the Space

Aspiring urban growers soon realized that not all vacant or blighted lots can be converted into a garden or a farm, despite their abundance in the city, which had already been experiencing economic shrinkage decades prior. Some of these lots were not appropriate for growing food due to poor lot conditions, while others lacked pre-existing water access. Even putting aside questions of access—legal or otherwise—prospective urban growers learned that not all lots were made equally. As they began their search for a space to implement their aspirations for urban cultivation, the growers encountered structural challenges that stood in the way of quickly enacting the changes they had envisioned. These initial barriers reflected the novelty of what they were trying to do—to start a larger scale food production in the city—because the spaces that appeared available were not accessible or not in the ideal condition or location for their projects. But the barriers also reflected the lack of pre-planning among the prospective growers, who tended to jump into action without fully anticipating the potential challenges.

One of the common drawbacks of the blighted properties was precisely that owners had not maintained them, physically or legally. The flooding caused by the levee breach in 2005 exposed a large area of the city to harmful materials, especially lead.[2] Dumping of toxic materials onto vacant lots during the recovery process further damaged the soil quality of these lots. But Black neighborhoods in the city had already been exposed to heightened levels of pollution due to their proximity to the industrial facilities along the Mississippi River, which itself carried agricultural runoff from the Midwest and the South. In short, the available vacant lots in the city were likely to contain high levels of toxic contaminants, making them nonideal for growing food for consumption. Ariel, who was starting a cultivation at her home in the Pigeon Town neighborhood, hoped to turn it into a food forest that would serve as an educational space for the community. In the process of preparing to start a new garden, she faced the challenges of contaminated soil:

My soil is actually highly contaminated, unfortunately. There's a significant amount of concrete, almost three fourths or more of the property is concrete. . . . I'm learning more about cut flowers, because I felt like I do have this resource of this little land that's native soil, because it is contaminated and there's so much glass and other debris from an automotive shop that was in the back of the property originally, which is where the contamination came from. And you know, I could still use it. I was thinking about hauling all the dirt away, like, taking out a foot and half of soil, and moving it. It's just very, very expensive to do all these things. So the next step is to do cut flowers in that section. And then probably just to focus on building raised beds on the property since there is so much concrete. It will cost at least $2,000 to remove a portion of the concrete. So you know it's just kinda like one step at a time, figuring it out.

Contaminated soil was not suitable for cultivation projects, not only because of the time and resources needed to clear a space, but also because of growers' concerns about health impacts. Ariel's lot posed further challenges in the presence of hazardous debris in general, from concrete to glass, like most other abandoned properties found across the city. Yet the growers pressed on, seeking solutions.

A common solution to contaminated soil was to use barriers to prevent ground toxins from seeping into the topsoil by constructing raised beds or planting in containers, as Ariel described. Jeanette tested the soil of one of the new gardens she was starting and had "no intention of planting any food items that would uptake lead and arsenic, which are two that everyone is very concerned about." When the test came back and the numbers were "really off the chart," she set out to grow in containers as she undertook an elaborate soil remediation process that involved spreading "five truck-loads of wood chips" over plastic, covering it with cardboard, covering *that* with topsoil, and then planting in the top layer. Other growers settled on growing things that would not take up the metal from the ground, or focusing on growing nonedible plants, such as flowers. Using sunflowers to remove lead from the soil was a very common practice during the recovery period, though some debates persist on its efficacy in pollution research.[3] A few, mostly younger new growers referenced research that showed that the highest risk of lead contamination came from direct human contact with soil and not from handling or ingesting plants that may have taken up the metal. They argued that raised beds were an unnecessary step, unless the grower wanted them for aesthetic reasons. Regardless of the scientific research, however, concerns about soil contamination and its impact on food production remained a major concern for most growers as they initiated their cultivation projects.

In addition to soil quality, an attribute that growers prioritized in identifying possible cultivation space was access to a physical water outlet or a working meter. In some cases, growers were able to make arrangements with next-door neighbors to use their water either for a fee, in exchange for bartered food, or for free. A grower who began his urban farm in 2012 on a lot he was leasing from Habitat for Humanity's HUG program with a couple of friends observed that he had selected the lot on the basis of its location ("upriver side of the industrial canal," close to where he lived) and the condition ("It was in the full sun most of the day, and had a nice palm tree in the back of the lot"). One drawback, however, was that the lot lacked its own water access. He was able to arrange to use a neighbor's water in exchange for payment and a share of the produce from the garden:

> So there's only one occupied lot immediately adjacent to the property, and they've been pretty supportive since the beginning. Actually, we've been paying them monthly to use their water rather than hooking up to the city water. It just ended up being a better deal for us and a better deal for them. And yeah, you know, we give them produce from time to time. Basically, we just tell them to go pick whatever they want, whenever they want, and, yeah, it's been a pretty good relationship.

Making such an arrangement was a "better deal" than trying to establish a formal water access on site, because it avoided dealing with the city's Sewerage and Water Board.

The Sewerage and Water Board was mentioned more than any other city agencies in my interviews, as the growers aired their grievances about dealing with bureaucracy. The agency's poor recordkeeping created confusion, and multiple calls to its office were often necessary to set the record straight about who was on an account and what was owed. For example, the requirement for agricultural water use of a "backflow preventer," a mechanism that prevents water from flowing back into the water source, had to be completed with a certified, specialized plumber. The estimates of this procedure ranged from a few hundred to more than a thousand dollars, and some growers questioned if this requirement even applied to their space.[4] One grower described the Sewerage and Water Board as "completely dishonest, lazy, totally disorganized," after spending months trying to correct the official records that their gardening site was not, in fact, a seven-unit apartment complex, and thus, the agency should not be charging them an additional fee associated with a multifamily residential unit. The incompetency of the agency worked as an advantage for at least one grower, whose meter was paved over, meaning the Board could not charge them. But in most cases dealing with the Sewerage and Water Board cost the growers time and money, which delayed the start of their projects. This was why some opted to make informal agreements for water access rather than establishing legal water access.

While some plots were more desirable than others, most growers maintained their commitment to their vision of urban cultivation and kept looking until they found a lot that met their needs. But then came the next challenge in the process—establishing access to it.

Establishing Land Access

Even for the spaces that met the growers' preferences in terms of location and quality, land access remained the most consistent hurdle to starting a new cultivation project. Figuring out who owned the lots and getting in touch with the owners took some investigative work or reliance on informative neighbors. Negotiating access, especially when the growers tried to purchase the lot, became a lengthy, drawn-out process, often with little result. Whenever growers faced these challenges, they tried to come up with solutions that allowed them to gain access more quickly. In many cases such solutions resulted in informal and precarious lease agreements, rather than transformative and sustainable changes that would produce more secure and formalized land access for urban cultivation.

Identifying potentially suitable lots required several strategies. For publicly owned lots, there was an option to search on websites managed by the City of New Orleans and NORA, though search functions on these websites did not become available until 2012, for BlightStatus managed by the city, and 2014, for NORA. The information on these websites was not always up-to-date and did not include privately owned vacant lots. Ultimately, growers usually needed to see a lot in person to assess its conditions before deciding to pursue access.

One of the initial prospects of ownership was to obtain publicly managed lots, frequently described as *adjudicated* properties in Louisiana, through auctions periodically held by the city. These properties had been acquired by the city due to failure to pay property tax or other violations, and the new owner would be expected to inherit those costs. The same issue affected private land purchase deals. Zach once tried to purchase a lot near his home, where he had started his original cultivation space, and gave up the process when he realized that it would cost significantly more to acquire the new lot due to the back taxes owed by its previous owners:

> So this lot behind me, taxes have not been paid since 1993. So the land has been adjudicated to the city, but it is still under the name of the people. And sometimes you can get a blight lien. Well, we've kind of shot ourselves in the foot, because we've cleaned up the area, you know. I'm trying to acquire that land. I've talked to the city councilwoman, I talked to other people. [The property] is essentially nobody's property, because the LLC that owned it is gone. The land value is like $9,200. But the back taxes are $18,000.

The state's succession law created another hurdle to those attempting to purchase a lot. In case of intestate succession, when the deceased did not leave an explicit will, everyone who is on the deed must sign off on the sale of the property.[5] This requirement resulted in complications whenever there were disagreements within the family, as there often were.[6] In 2014, David described one of his experiences of trying to purchase a private lot that stalled and eventually failed:

> I tried to buy a house that was beside my first lot. Six siblings, four wanted to sell, one didn't, and one was undecided. So [I] never got that bought after a year, you know? One of the colleges was helping with the succession issue. And they said that now they were six years behind with all the cases that they've had. Every case takes two and a half years or some enormous amount of time, because they have to do a whole family tree and they usually go back six generations before they find out who actually owns it [before they] try to sort it out.

Stories like this one were very common among growers who began pursuing the purchase of a privately owned lot only to realize that it was a dead end, making it impossible to gain long-term land access through purchase.

Even when growers could pursue property acquisition without these two issues becoming obstacles, securing access through purchase remained a complicated process. Daniel, who held a leadership position at VEGGI Cooperative, a cooperative for Vietnamese immigrant growers in New Orleans East, compared the differences in negotiations with private owners who still lived nearby with owners who were in another state or even overseas. He explained:

> So with the community member, it was really just asking folks around the community if they knew who that person was and if they could get us in contact with them. And basically figuring out if we could schedule a meeting with them. . . . That landowner knew of what we were doing, so we just had a conversation about what we were trying to do, what we've done so far, and kind of the potential benefits to the community, what we are going to be looking to do and what we aren't going to do, kind of what they want us to do or what they don't want us to do on that property. And then trying to work out a reasonable timeline that fits their timeline as well, with the mission and visions for that property.

The property owners in the area were aware of VEGGI's work in the community and were more open to negotiating for a lease. By comparison, Daniel recounted that out-of-town owners not familiar with their work tended to be less flexible in providing affordable and preferable lease terms for the nonprofit organization, which worked to create a source of income for Vietnamese immigrant growers. The inflated or at least hopeful expectations about property values among property owners was not limited to out-of-town owners. Others were open to long-term leasing but resisted the idea of selling a property that had been in the family for multiple generations. Given these constraints, leasing the lots proved much easier than trying to purchase them, especially for privately owned properties.

Historically, cases of community-led guerrilla gardening projects benefited from ample social resources despite limited economic resources. But even urban cultivation projects that aspired to be a part of the post-disaster community rebuilding process did not always find enthusiastic or supportive neighbors. During the recovery period, returning long-term residents by and large did not see urban gardening as the priority, as their focus was on rebuilding their own homes or major institutions such as schools, transportation systems, and hospitals. As the city entered the transitional period, growers found that some nearby residents showed interest in their projects or even actively supported them, while others remained indifferent or disliked the idea of having cultivation projects in the neighborhood. One nonprofit organization leader recalled a meeting he had attended during the transitional period in the Upper Ninth neighborhood, where he witnessed residents at the meeting vocally expressing opposition to the idea of urban agriculture in their community:

> They were, like, adamantly opposed to any urban gardens, like they were going to fight to the tooth to not let any rezoning occur. . . . There was no way on earth that they were going to have—it was going to, it was kind of an "over our dead bodies" kind of [thing], their reaction.

He added that when his organization decided to build a fence around its gardens, a neighbor expressed disapproval, commenting that the fences would be an eyesore. Long-term residents' resistance to urban cultivation is likely a reflection of their resistance to attempts to change the larger fabric of the community, not specifically about their objection to gardens or gardening in particular. The fact that many of the new urban growers were white, middle-class newcomers could have added racial and class layers to long-term residents' skepticism toward the broader implication of these projects.

Most growers tried to reach out to nearby residents to inform them about the project plans and seek their blessing, even though this was not required, and the neighbors who did not support the projects were usually not openly hostile. Jerry observed the lack of enthusiasm and support from the neighbors for his youth-focused urban agriculture project, ACRY (A Community of Renewed Youth), as stemming from the economic struggles prevalent in the community:

> I attempted to reach the community. Actually, I tried to talk to the peo-ple in the hopes of getting them to help with the water issue. But unfor-tunately, poverty has a strange effect on our species. It doesn't allow for us to see the bigger picture. We get very tunnel vision when [we] are

deprived of the basic necessities of life. So therefore, you can't see, or comprehend, or even have the time or energy to even think about how, you know, something as small as a garden could actually have a huge effect on the people and the surrounding areas, you know. They just didn't have energy to [extend]. They weren't opposed to it. They smiled about it, but that was about as far as it went, though. No personal connection to it at all, even though it was in their neighborhood.

A Black New Orleans-native himself, Jerry attributed the disconnect to a generational distancing from agriculture and the general lack of knowledge of "where your food comes from" among the youth coming up in the city today, and even within his own generation. So when he reached out for support, the responses often came from individuals who did not live in the community, despite his hopes of wanting to engage the people whom he felt were most in need of the knowledge and the experiences of how to grow one's own food. Growers reported that interpersonal expressions of reservation or hostility toward urban cultivation were rare, though they suspected that some in the community would have preferred to see their neighbors return to rebuild homes rather than a newcomer start a garden.

Building rapport over time by putting in sweat equity paid off. Our School at Blair Grocery took its name from its predecessor, a former family-owned grocery store that had been closed since long before the 2005 storms. When Turner was exploring ideas for how he was going to continue his work in the Lower Ninth Ward, a neighbor he befriended initially connected him to the family that owned the property. The family was not ready to sell it but was willing to lease it, on the condition that the organization pay for the property tax and worked to rebuild the structure, which had been severely damaged during the 2005 flooding. After the first few years, the eight children of the original owners told him that they were open to a ten-year lease on the property, because they saw that the property was being managed and thought their mother would have wanted to see the place put to use for "something good," even if it was not going to be rebuilt as a grocery store. This somewhat serendipitous case of access to space exemplifies the kind of neighbor-facilitated access to the land that became an effective solution when it was difficult to identify or find contact information for property owners.

The long-term residents' tight-knit connection to their former neighbors and the common practice of extended relatives living in proximity to each other made it possible for interested growers to get in touch with a relative of the owners. But the growers also had to gain residents' trust in order to tap into the community's social capital.

Start-up Costs

Another hurdle for a new cultivation project was the cost of starting up, including but not limited to land-access fees. Jeanette described one of her gardens as "an investment in excess of 100,000 dollars," recognizing that "the average working person doesn't have those kinds of resources to put into a hobby like this." Pamela noted that the start-up cost is something that most novice growers do not fully consider when they conceptualize a new garden or farm. She kept meticulous notes on every expense that each of her projects had incurred, and would advise aspiring new growers on the "true cost of start-up and ongoing maintenance," because she had observed that skirting these initial costs was often the key in projects not ultimately materializing.

As Jeanette recognized, not every aspiring grower had access to start-up funds. Those who were able to self-finance were mostly retirees, many of them with teachers' pensions, who could afford to purchase a lot or pay for materials and services on their own. But even those who self-financed their initial projects eventually sought alternative sources of funding for their subsequent projects, often through private foundations in the form of various grants. Rob, who moved to New Orleans in 2009 and joined Our School at Blair Grocery and later became co-director alongside Turner, described the funding that the organization had secured:

> [At the beginning,] that was all Turner's funding from his pension. But I guess about eight months after we started, we got a grant from the Greater New Orleans Foundation. And that helped us get going a little bit more and then after that we got a grant from the Kellogg Foundation, a couple private foundations, and eventually, I guess about two years after starting, we ended up getting awarded a grant from the USDA. . . . Louisiana Delta Service Corps helped fund staff that we had at the time. That's what a lot

of the grants went towards, paying the matching funds for the LDSC people to work here. We took advantage of a lot of [grants] here and there.

The range and types of funding sources Rob mentioned were typical of many cultivation projects in the city. Some of the grants were aimed at a specific aspect of the cultivation project, such as youth engagement or nutrition programs, depending on the foundations' priorities. Around 2008–2010, the growers found a resonance between urban cultivation and foundations' interests in the social issues of childhood obesity, community health, and urban sustainability; as a result, the growers began applying for grants from local foundations such as the Greater New Orleans Foundation and national organizations such as the W.K. Kellogg Foundation. While they did not provide direct funding for cultivation projects, programs such as AmeriCorps and Louisiana Delta Service Corps provided paid interns and personnel that assisted in the design and execution of the programming, especially during the first half of the decade. But those grants and programs were primarily available to pre-established projects, and in a few cases they continued to fund the same projects over multiple cycles.

The founding of several online crowdsourcing platforms around 2009–2010, including Kickstarter, Indiegogo, and GoFundMe, created a new avenue for growers to raise funds without relying on large foundations. These platforms quickly became widely used by a variety of entrepreneurs, nonprofit organizations, and individuals to rely on small-scale funders to complete specific projects. Use of these online fundraising platforms required some technological familiarity and access to digital resources to set up, so it was mostly younger growers who relied on this type of fundraising, regardless of their urban cultivation aspirations. The funds being raised for urban cultivation projects were relatively small, usually from $1,000 to $5,000, and were typically used for installing specific pieces of garden infrastructure or for general start-up costs. In exchange, the growers typically offered varying shares of fresh produce to their supporters, depending on the scale of their investment.

Before they were able to start seeking larger grants, the growers explored various funding sources, from crowdsourcing and fundraising events to accepting in-kind donations and volunteers to deflect the cost of labor and soil or other essential materials. Jenga's Backyard

Gardeners Network fundraised with a crawfish boil and a container gardening workshop, and cut neighbors' grass for donations in order to raise enough money to purchase the Guerrilla Garden lot from NORA. In 2009, Backyard Gardeners Network was awarded $12,000 from In Good Company, which finally helped the organization to purchase the lot that it had initially begun tending without formal access before gaining formal access through NORA. In addition to the financial support, In Good Company, an alliance of "environmentally and socially conscious" food companies that includes CLIF Bar and Annie's, organized a group of volunteers from across the United States to visit New Orleans to help with the initial setting up of the garden.

Shifting Land-Access Opportunities

Over the decade following Hurricane Katrina, land-access opportunities for urban cultivation projects changed, reflecting the economic and social transformation of the city. At first glance, it might have seemed easier to start projects right after the storms: the recovery period presented a vast stock of vacant and blighted lots, and interests in urban cultivation expanded as the city transitioned toward redevelopment. Nevertheless, finding and accessing space for a new garden or a farm was not necessarily easy for growers at that time, partly because the city was not ready for this type of prefigurative urbanism. As the city continued to transition toward redevelopment, new interest in and support for urban cultivation began to change the opportunity structure. Yet these new opportunities also posed a new set of challenges for growers, especially the risk of being exploited by those who wished to use urban cultivation as an affordable form of land management until property values increased. Rather than rejecting these new possibilities altogether, growers attempted to engage selectively with them while maintaining autonomy over their projects. Some projects successfully managed these partnerships and collaborations, while others experienced failures or at least some setbacks.

Recovery Period—New and Exploratory

During the recovery period, the vast stock of vacant and blighted properties could have appeared to many as the best opportunity for growers to gain access to spaces and start cultivating. They could have bypassed the formal mechanisms of access and just taken over unused land to start growing, and asked not permission first but forgiveness later. This type of gardening would qualify as a form of "guerrilla urbanism"[7] and does happen in many cities. But this was rare in New Orleans during the recovery period, partly because there was still uncertainty about which lots were waiting to be rebuilt and which would remain blighted; finding information about these lots was not an easy task, and trying to formalize access was time-consuming and sometimes created more problems.

After 2009, when Jenga helped revive a former community garden site in the Lower Ninth Ward that had been vacant for a long time before Katrina, she and local residents began growing on the lot with no initial intention to seek permission on the lot and called it "Guerrilla Garden." When they got a donation of trees to plant from the Fruit Tree Planting Foundation, the nonprofit organization required that they show long-term lease or ownership as a condition for the trees. This was when Jenga and her collaborators realized they need a permit to plant the dozen trees, and to get the permit they needed to prove ownership or a long-term lease with the owner. When they began to look into the lot's owner, they learned that it was owned by NORA, which not only denied their request to plant the trees but told them that they were trespassing and needed to cease their use of the space while a usage agreement was being worked out. After more than a year of efforts, Jenga was able to raise sufficient funds for the down payment and negotiated a unique arrangement with NORA whereby the Backyard Gardeners Network could purchase the property with a ten-year, forgivable mortgage for the remainder of the value of the property, on the condition that the property be maintained as a community garden for the duration.

In the end, this was a victory for Jenga and the group of Holy Cross neighborhood residents in the Lower Ninth Ward, whose work on the lot acquisition inadvertently became an organizing process. But this was an exception among the land acquisition attempts and challenges across the decade. In general, when growers' attempts to access a given lot failed, they moved on to seek another space, rather than using the challenge as an opportunity for mobilization. For Jenga, who came to the idea of gardening through her work in community rebuilding, moving to another location to start a garden did not make sense. When NORA exercised its ownership authority over access to a space the community had been tending, albeit without legal permission, this resulted in Jenga and the community members' decision to mobilize around this specific challenge, formalizing the ownership of the space that they had already improved and were using for the community.

Across town in Central City, in 2008–09, Pamela began planting sunflowers for lead remediation on an empty lot on Oretha Castle Haley (O. C. Haley) Boulevard with Dillard University chemistry professor Dr. Lovell Agwaramgbo and Danielle Purifoy, an intern from the City of New Orleans. Café Reconcile, a local nonprofit that runs a restaurant where at-risk youth receive job training in the hospitality industry, owned the property. Once the remediation project was successfully completed and a small garden was installed, Pamela approached the director of Café Reconcile and negotiated a lease to formalize the development of Sun Harvest Kitchen Garden. The lease was at no cost, with the stipulation that Pamela would maintain the lot and develop the garden infrastructure. As a long-term, multi-generation New Orleanian grower, she saw this as an opportunity "to demonstrate how to properly put in infrastructure for an urban property like that," meaning an abandoned property with no pre-existing cultivation infrastructure, including water:

> The lot is actually 65 by 85 square feet. It's a corner property. And what I did was first look into connecting the water spigot. So, the Sewerage and Water Board discovered that there had been a building there many, many, many years ago, but the meter was no longer there. They researched to find where the connection should be, but I actually had to pay for a meter and the water connection. The type of water connection I had to get was for agricultural use with a back-flow preventer. It saves the $24 fee for garbage pickup. So, whenever I had debris there, I had to remove it. I keep meticulous records of costs for everything. As I recall, the meter installation was somewhere between $750 and $900 to get. You have to hire a licensed plumber to file the permit, put in the water line, and install the back-flow preventer.

The cost of water access installation became a major obstacle for many new projects, in addition to energy required to navigate the bureaucracy

Figure 3.1. Sun Harvest Kitchen Garden in preparation. Photo by Pamela Broom.

to have the proper plumbing work completed. Thanks to her work at New Orleans Food and Farm Network (NOFFN), Pamela had a pre-existing relationship with and knowledge of some of the city agencies, and she attributed her successful navigation of the water access process to her familiarity with city bureaucracy.

While they waited for the water to be set up, they had a "hand-shake agreement" with the neighboring business to use their water for the sunflower remediation project, which also required a significant outlay on Pamela's part. Pamela had decided, after the remediation project, to implement an extensive soil-building process so that the site could grow produce for the Café's use. The organization agreed, so after the successful eighteenth-month cultivation of sunflowers and favorable lead-removal results, Pamela instituted a weed abatement plan. She and her daughter Lucinda Rose "covered the entire lot in cardboard, heavy wood mulch, and we began building soil on the top of the whole space" before constructing raised beds and implementing garden designs (see Figure 3.1).

The cost for the project continued to accrue as she began to design and build the garden, as she recalled:

So, after taking over the management of the garden, I put in a sort of decorative fence, split rail fencing along the front, the O. C. Haley side, and Erato Street side, and a wall from the business next door which left the back of the garden open. The cost of that fencing was around $1,000. It served the purpose of beginning to define the space. About three months later, I put in a chain-link fence because I wanted to demonstrate that if you are going to grow for the marketplace you really have to be mindful of not having animals freely walking on the site. The chain-link fence was six feet high along the sides and rear, four feet high across the front with a gate, and cost $2,300. It offered an excellent space for trellising.

It was August of 2009 when Pamela was finally able to begin growing on the lot, nearly two years after she began preparing it, just as more aspiring growers were starting to conceptualize their cultivation projects. Even though her work in the neighborhood did not receive as much media coverage as those in the Lower Ninth Ward, Pamela's work on Sun Harvest Kitchen Garden foreshadowed a range of possibilities and constraints that many new endeavors would face as more growers attempted to start new urban cultivation projects.

Transitional Period—Expanding Opportunities with Some Limitations

As more aspiring growers attempted to enter urban agriculture during the transitional period, a fragile infrastructure began to emerge to support their efforts. The growers began to take advantage of various opportunities for land access, from collaboration with nonprofit organizations, to community networking, to inheriting spaces from other cultivation projects. Landowners were becoming more open to the idea of allowing urban cultivation on their properties, as they recognized that those properties might not be rebuilt or developed for a long time, but still preferred leasing to selling. The terms of agreement for leases tended to be informal and varied in the length of the lease. The growers generally found the private landowners to be hands-off; these owners did not meddle with growers' day-to-day operations as long as the lots were maintained, which seemed to be the primary, if not sole, motivation for them to lease spaces for cultivation projects in the first place.[8]

Ashley started Trouser House at her rented property on St. Claude Avenue in 2009. The property had a double shotgun structure, a common architecture in New Orleans with two narrow, shotgun homes sharing a wall in the middle, with a shared backyard space. She rented the entire structure, which was commercially zoned, on a "residential style lease," to live there while running a business that she envisioned as "a contemporary art and urban farming space." The project's mission encompassed the *community rebuilding*, *alternative careers*, and *social entrepreneurialism* aspirations. She recalled the leasing process to be almost too easy. The owners agreed to let her sublet a part of the building, and "minimum modifications were allowed without prior consent," which meant she could install a garden in the back. She reflected on the blasé nature of the original lease and described the property owners as "not very good landlords," but observed that their hands-off attitude "was actually attractive to [her]" because of the freedom it gave her to start her new cultivation project.

Trouser House began operating without a business permit, and this reflected the general attitudes in the areas that were starting to show signs of transitioning beyond recovery. Ashley described how the Bywater neighborhood was at the time:

> It was kind of interesting to be on this kind of edge of an area that was, you know, still largely blighted and mainly home to low-income folks. And then right across the street, places like Satsuma [Cafe] and these really kind of trendy [businesses] started popping up, and houses were getting renovated left, right, and center. When I first arrived, we basically took the fence down between the two houses, or between the two sides of the shotgun [house] and cleared out the yard. There was still a bunch of hurricane debris in the backyard. . . . [In 2009,] [Mayor] Nagin wasn't too interested in the Ninth Ward, and there was kind of this rogue sensibility across the neighborhood, and especially about running under-permitted spaces.

The Bywater neighborhood, where Ashley began her cultivation project, did not experience major flooding during the 2005 storms, due to its proximity to the Mississippi River levee, the elevated topography often described as the "sliver by the river." As a result, it was one of the first neighborhoods that began to undergo rapid commercial and residential transformation during the transitional period. Satsuma Cafe's opening in 2009 heralded the commercial renaissance in the neighborhood, while other areas such as the Lower Ninth Ward continued to show very few physical signs of recovery.

Trouser House rented spaces to tourists through Airbnb, which was a relatively new platform at the time, as part of its income stream. The idea was that the funds would be used toward arts programming, including the on-site gallery and workshops. The garden became a central part of the immersive experience for the guests, who were encouraged to harvest as needed during their stay. The individuals directly engaged with the project were younger artists, many of them post-Katrina transplants like Ashley herself. While getting Trouser House off the ground, Ashley taught permaculture at a West Bank middle school and was actively engaged with the neighborhood rebuilding efforts in the St. Claude area, attending city council meetings and advocating for local businesses.

The trouble began when she decided to obtain permits for her business. By 2010, with new mayor Mitch Landrieu's administration placing emphasis on regulation and blight reduction, she felt the need to get a proper permit. Ashley described her experience when she went to city hall to get a permit for the place:

> They told me that the permit that I needed would require me to completely change the structure of the house and do all this crazy stuff, because we were zoned commercial, but [the property] had not been used as a commercial space. So I needed to come up to code for that present year. But you know, the house was built in the 1890s, so it would have been near impossible to bring that space up to code.

To top it off, even if she could afford to make all the required renovations to the property to bring it up to code, she was told that the permit for the kind of business she wanted to operate did not exist. After four months of trying to get clarification on the permitting and not finding any solution, she decided to terminate Trouser House's operation rather than continue operating without a permit. After briefly working on a small, sustainable-farming project in rural Mississippi, she left New

Orleans in 2012 to take a new job in another state, just as the city was about to see the rapid expansion of urban cultivation.

There was increasing support of urban cultivation projects by local nonprofit organizations and universities during the transitional period. But often, this support was not well coordinated, and was promoted by individual employees or faculty members with personal investments in the idea of urban cultivation. Parkway Partners, which played a central role in the 1980s community gardening movement, was not formally providing extensive support beyond free seeds and access to its greenhouse. But one of its employees personally reached out to make calls to property owners to find spaces for growers. Half a dozen growers attributed her assistance as a starting point for their gaining access to a cultivation space on a privately owned lot.

The site establishment of Grow Dat Youth Farm in the City Park, a 1,300-acre public park, is an atypical case of a successful collaboration across multiple organizations for securing an unusually large and distinct instance of land access. Johanna Gillian, who used to be the community organizer of New Orleans Food and Farm Network, conceived of and founded Grow Dat Youth Farm project in 2011. It was inspired by successful youth leadership training programs such as the Food Project in Massachusetts and aimed to hire local high school students to work after school and for part of the summer. The idea was to engage the youth in hands-on agricultural skills and leadership training while also teaching them about food systems, the environment, health, and social justice. During the initial year of programming in 2009, Grow Dat Youth Farm operated on the newly opened Hollygrove Market and Farm, where Macon and another long-term local grower were working as mentor farmers, while its permanent site was being developed at the seven-acre space in City Park. It included the learning center and office spaces that were constructed out of repurposed shipping containers. The establishment of Grow Dat Youth Farm on this particular site involved collaboration between the City Park Improvement Association, Tulane University, and the Lagasse Foundation, which eventually pulled out of locating its own farming project on the adjacent site. The initial agreement was a "triangulated agreement," whereby City Park and Tulane University would have a cooperative endeavor agreement (CEA) for the use of the site and Tulane University would subcontract the space to Grow Dat Youth Farm.[9] Considering that Grow Dat Youth Farm was a new nonprofit organization and a new urban cultivation project, the lot was an anomaly for its size and location. Even though the program continued to expand over the years and would become fully established as one of the prime examples of successful urban cultivation in New Orleans, at the time the access to this unique site was likely not possible without investment and support from the city's largest private university, whose president reportedly indicated personal interests and investment in the project.

The transitional period saw a relatively successful expansion of existing projects, a few of which had begun during the recovery period. Growers used a variety of methods in identifying and expanding onto additional spaces. In general, the growers preferred prospective new lots that were near their previous cultivation projects, for the ease of daily commuting from site to site to tend soil and plants. For projects with *community rebuilding* aspirations, the cultivation project had to be located within the specific community where they wished to work, which geographically limited their search for potential new lots. During this period, there was a concentrated emergence of cultivation sites in the Lower Ninth Ward, both because many growers sought to work in a community that was experiencing a slower rate of recovery, and because more landowners were coming to terms with the real possibility that they might not be able to return to rebuild.

As more aspiring growers attempted to start their own projects in New Orleans, existing growers faced new challenges, particularly competition over available space. After initially starting Capstone in 2009 while still living in volunteer and community housing, David gained access to twelve additional lots in the Lower Ninth Ward from another cultivation project that was being run by a nonprofit organization; this acquisition quickly expanded his project's land access. While the scale of this acquisition makes it a unique case, some of the challenges he faced exemplify common challenges in land access during this time. David recalled the initial process of establishing a new lease for these lots with the organization as relatively straightforward; the process occurred mostly through personal conversation, and the owner's personal trust in his work in the community played an important role. Once David and the volunteers spent three months clearing the land, another grower, who had initially informally taken over the use of the lot from the organization but did not act upon it, expressed interest in the space. In the end, the property owner decided to lease it to David because she told him she respected him and his work in the community. Fortunately, it worked out for him, but David risked not gaining access to the lots after spending hundreds of hours readying them, due to the precariousness of the understanding under which he began working on the space.

Even successfully established land lease agreements typically prioritized the needs of the owner. When I asked him in 2014 about the nature of his lease with the owner for the twelve lots, David responded that it was for an "indefinite" term, clarifying that there were no specific terms under which the lease could be dissolved. He continued:

> And it's done with the understanding that at some point in time that they have grandkids who might want to build there, that type of thing. You know, she said, "We don't want any trees planted there, 'cause we don't know what the grandkids will wanna do." Now, the grandkids she's talking about are like *this* big [indicating they were still young].

The precariousness of the lease and the owner's general preference that the property be maintained in ways that would leave the future use of the lot open-ended exemplifies the selective, reserved support for cultivation projects as placeholders rather than as permanent establishments.

Redevelopment Period—Come and Farm It!

By 2012, the number of cultivation sites began increasing significantly, more than doubling from 37 lots in 2011 to 88 lots in 2015. The increased number of people entering urban cultivation meant that new growers were able to start new projects quickly, while existing growers likewise were able to expand their existing projects quickly. During this time, more municipal organizations and nonprofits, including NORA and Habitat for Humanity, began leasing properties under their ownership. New Orleans Food and Farm Network's new website attempted to facilitate access to information on these newly available spaces. But despite the increased interest in urban cultivation, these property-owning institutions' interests in urban cultivation did not align with the growers' own aspirations. The institutions treated urban cultivation as a cost-effective and publicly appealing way to outsource the management of their properties, even if the individual staff members assigned to manage the programs were personally supportive of the growers. The growers selectively and cautiously responded to these new opportunities for land access, even as they recognized that the external interests were primarily in land management rather than in supporting urban cultivation itself.

Before Nicole signed a sublease with VEGGI Cooperative in New Orleans East in 2012, she was "on a mission" to find a large space for her cultivation project. She initially began her search in North Shore across Lake Pontchartrain, more than an hour away from downtown New Orleans, but she eventually realized that the commute was not feasible, especially if she wanted to have growing partners from the city. She then turned to the Lower Ninth Ward and began looking for NORA properties, which were easily identifiable because they were always mowed and had their addresses spray-painted in yellow on the curb. She drove around the neighborhood, making note of a few sets of contiguous, vacant lots that were owned by NORA, as well as adjacent lots with a "missing tooth" lot in the middle that was owned by a private owner. Using the Orleans Parish Assessor's website, she found out who these owners were, tracked them down, and asked if they would be willing to sell their lot. Some of them said they were.

At this point, Nicole had applied to lease a lot through NORA's Alternative Land Use program. Though she described NORA as having "good intentions," her attempt to access the lots primarily resulted in frustration:

> So I filled out an application. And my final application was like eleven pages of, you know, like my business plan. And then they asked for, you know, a pretty specific plan, of not just your business plan but also like, your commitments for what you would do with the property, as far as maintaining it and buying [it] from the community. And so I got letters of support from the people in the community. . . . So I invested myself in forming relationships in the community and sort of like making

some commitments about my intentions. It was a very lengthy process. Talking to the neighbors, forming relationships, getting that support from them, and then submitting my application to NORA.

The agency confirmed that they had received the application but then gave no further response for weeks. When they finally did, they told her that the lots she had requested were not the most "suitable properties" for her project. She remembered that the agency instead offered her several non-contiguous properties:

> And so we had a meeting, and you know I appreciate them taking the time to have a meeting with me, just one person. But I was trying to explain to them, for a farmer, you can't be fencing off this, and paying to have, you know, because you have to have water. And it's like, I know from experience, from the land I'm on now, it costs $10,000 to put in a water line. You can't do that at five different locations. Build a shed on each location? It's just not feasible. And so we talked about that. They did get back to me again, and they offered me [some other lots]. I mean, they just, I think they're well-intentioned, but I think they just kind of walked away, and I just realized that they just weren't in the same place I was at. You know, I was really interested in moving forward, and they had other things on their plate.

Nicole suspected that the agency was withholding some of the larger lots for its own projects, and thus was not willing to lease or sell those to growers. After giving up on lease through NORA, she ultimately landed a sublease from VEGGI Farmers Cooperative in New Orleans East, with whom she had built a relationship when they were both fellows at Propeller, a local incubator that founded in 2011. Her journey of seeking space for cultivation in multiple locations and through multiple paths, including a failed attempt to lease along the way, exemplifies the tenacity with which growers starting a new cultivation project around 2012 needed to navigate the task of establishing land access.

In 2014, NORA unveiled the Growing Green program to address the ongoing demand for land for urban cultivation. The program would charge an interested grower $250 a year to grow on one of the agency's 2,500 vacant lots. An earlier program, Growing Home, had streamlined the purchase process for owners of adjacent lots, but few aspiring urban growers met this condition. The Growing Green program expanded the Growing Home program by removing the "shared property line" eligibility requirement, opening up an opportunity for anyone wishing to use the space for a variety of green purposes, including gardening and farming. The original description of the program posted on the agency's website described its four objectives: improve neighborhood stability, foster neighborhood safety and sustainability, make fresh produce available, and/or promote a general sense of community.[10] This was part of the agency's intensified effort to reintroduce to market the properties that it acquired through the Road Home program, hoping to reduce the costs associated with managing hundreds of vacant lots across the city.

Growers who had been looking to own their own space were particularly excited about the agency's position on tenancy. Though approved applicants would start off with a "standard one-year lease," renewals were possible. According to the website, "Participants who manage a successful project for at least two consecutive years may be offered the opportunity to purchase the property."[11] This did not mean, however, that the agency would be willing to sell or lease its land to any interested applicant. The agency's vision for these lots was not just to improve their value through beautification and maintenance, but to contribute to redevelopment of the area. In short, the agency considered urban cultivation a valuable use of their property, so long as it met the agency's vision and expectations of redevelopment.

One grower that we interviewed found working with NORA straightforward and described the agency as "the most efficient organization I ever worked with," but this sentiment was an exception rather than the norm. The majority of the growers expressed their frustrations with the bureaucratic process of establishing a lease, as Nicole's story exemplifies, and voiced skepticism about whether they would ever be able to purchase the land. When compared to mostly hands-off, informal lease agreements with private owners, the rigid and formal processes of NORA struck many growers as off-putting, especially when negotiations could take many months, during which the terms of agreement kept changing. A grower who had at one point explored leasing or purchasing a lot from NORA told us that she would not touch it "with the tip of a ten-foot pole" and listed her concerns with the way the agency's programs favored its own priorities over the growers:

> NORA will only sell two lots at a time to growers, and it takes a lot of loopholes to even acquire the lots. But what ends up happening is lots are acquired in like a checkerboard, so they're not contiguous space to grow. Even if they sell it to you they retain mineral rights to the space. They also retain the right to revoke the sold land if the properties don't fit their criteria of what "looks appropriate," but they don't issue what exactly the criteria of what "looks appropriate" means. And also upon purchasing the land, you revoke your right to litigate against them. So if they decide they want to take their land back away from you, you cannot sue them. You have no right to say anything about it. So those are my main issues with it. There are some smaller issues, but I really have been hesitant to go through NORA as it stands to acquire properties.

Her description of the constraints and expectation of the sales agreements was confirmed by the documents she had shared with us later via email, and my conversation with the NORA representatives generally validated these points. The agency had its own agenda, and urban cultivation was one of the many potential uses for the large stock of properties that it managed.[12] Interestingly, this grower had also experienced multiple failed attempts in accessing privately owned lots, but she did not describe these to be systematic issues of intent or an attempt to co-opt urban cultivation into redevelopment efforts.

For a grower, a lost planting season means a lost harvest season. But it was far more than time that underlay the growers' general hesitance to apply for NORA's Growing Green programs for their cultivation projects. It is likely that at least a part of growers' skepticism toward NORA mirrored the shared public sentiment toward the government, in the context of the post-Katrina failure of state and local agencies to protect the interests of everyday citizens, especially marginalized populations.[13] The NORA representatives whom I interviewed expressed sympathy toward the "confusion" caused by the "miscommunication" that they saw as partially resulting from the agency's own lack of access to full authority over the lots, as well as the 2012 change in leadership that led to a subsequent reorganization of the internal information management process. Another

NORA representative also revealed that some of the lots that may appear as "available" on their publicly searchable list at the time were in fact being internally "reserved" for other purposes, but for a long time, this information was not communicated across the agency, leaving applicants perplexed and misled. Multiple growers we interviewed shared similarly confusing patterns of lengthy or failed attempts to lease lots from NORA, indicating that the agency was only interested in allowing urban cultivation on lots that it deemed as lacking potential for development, as NORA did not view gardening or farming as sufficiently valuable on its own. To an extent, this made sense for an agency that did not want to see the lots it leased or sold left fallow, but such cautiousness and lack of transparency in the process contributed to the widely shared skepticism toward NORA among New Orleans growers by 2015.

The growers' common suspicion about the intent of NORA's Growing Green program contrasted with the way they viewed Habitat for Humanity's Habitat Urban Garden (HUG) program, which launched in 2012. Similarly to NORA, Habitat for Humanity made available the list of properties that it had acquired, either through sale or donation, especially those it had no plans of converting to homes anytime soon. Margee, who is white and grew up in New Orleans and North Shore, studied horticulture at Louisiana State University, and while in college worked on developing a network of community gardens in Baton Rouge, Louisiana. After returning to New Orleans in 2011, she got involved in several gardening projects across the city while working on advocacy for locally grown flower businesses through the New Orleans Flower Collective. When she started looking for a cultivation space of her own, she unsuccessfully tried many tactics that were commonly being used by growers at the time, including attempting to secure a NORA lease and cold-calling owners of blighted properties that she had identified on the Assessor's website. She belonged to the first set of growers to sign leases with the HUG program. After trying many approaches to accessing land, Margee found the HUG process to be much more straightforward:

> They gave us a list of the properties back when they were kind of pioneering this project. I drove around for probably four days looking at all of the properties, assessing them for the things that I wanted. I had originally picked [one lot], and when I went in to sign the lease they really wanted

me to take [the lot on] Mandeville [Street]. I didn't want it because it was a corner lot, and it was unfenced. And so, as part of the bargain they fenced it for me. I took [the lot on] Mandeville 'cause they gave me a six foot chain link fence, which is just the most cost prohibitive thing for a gardener in this city.

After that, she signed a five-year lease at the cost of one dollar a year, and, two years after the start of the lease, at the time of our interview, she didn't think the organization even cashed those checks. Overall, other growers who leased from HUG similarly expressed how straightforward the process was, especially when compared to NORA's. Habitat for Humanity did not meddle with the specifics of grower practices, except for the requirement that they maintain overgrowth and not build a permanent structure on the lot. The growers did not always get the exact lots they wanted off the list, and in some cases Habitat for Humanity suggested a year-to-year lease rather than a full five-year lease from the beginning, signaling that the nonprofit organization had priorities and plans for some of their lots, similarly to NORA. Regardless, its reputation among the local growers remained generally positive when compared to the state agency.

When I spoke with two HUG representatives in 2015, they each confirmed that the selection of the available lots was based on the organization's assessment of which areas were least likely to be preferred by the potential first-time homebuyers with whom they work. When I noted that a dollar a year for five years seemed "generous," one of the representatives responded, "I mean, it's sort of mutually beneficial. It's sort of low-cost access to land, but, I mean, these are a lot of lots. It costs us to, you know, it costs us money to maintain, and now they're responsible for maintaining it, so." While HUG's use of urban cultivation as a cost-effective form of land management was the same as NORA's in principle, the growers gave private landowners, including individuals and organizations like HUG, a pass for their relatively hands-off approach to leases, possibly because it signaled trust in growers' intentions for the lots.

In 2012, David purchased two NORA lots through Propeller's Pitch NOLA contest called "Lots of Progress." The contest was set up in collaboration with NORA as a "live pitch competition event" that challenged "anyone in New Orleans to propose a creative, viable, sustainable and scalable use of a vacant or blighted property that would have positive impact on the community."[14] Yet after winning the contest, David found the process of finalizing the sales no more straightforward than when he was directly negotiating with NORA. He reflected on the process as having a big "learning curve," and felt that formalizing access to the lot "would've been enough to drive me over the edge" if he did not already have prior experience in real estate transactions. While well-intended, Propeller's program did not provide guidance to facilitate the land access negotiation with NORA after the competition, thus leaving the growers to finalize access on their own.

Propeller was not the only nonprofit organization stepping up its efforts to facilitate land access for growers during the redevelopment period. In 2014, the New Orleans Food and Farm Network developed the Living Lots NOLA website, as a part of its FarmCity Toolbox project. The website was modeled after the "596Acres" project in New York City, which geographically mapped vacant lots available for use by the public, including urban growers, by providing information on the owners and identifying the pathway to gain lot access based on the type of owner (see Figure 3.2). Living Lots NOLA synthesized the information of several agencies including NORA, the Housing Authority of New Orleans, Habitat for Humanity, and other privately owned properties whose owners agreed to be put on the list. The aim was to democratize access to the information and to facilitate land access for anyone interested in finding and establishing access to spaces for urban cultivation.

Despite the good intention of the initiative and attempts to distribute the information widely among the urban grower network, the growers mostly remained unaware of the service or found the information ultimately unhelpful. One grower reported finding one of their lots through Living Lots NOLA, but others noted that after searching for lots on the website, they eventually found their preferred lot by driving or biking around neighborhoods or going through a specific agency or organization's list. The website was active for about a year before it stopped operating. The failure of the Living Lots NOLA website to facilitate land access for growers did not necessarily stem from the usual factors, such as a digital divide or social capital segregation, because it did not expand

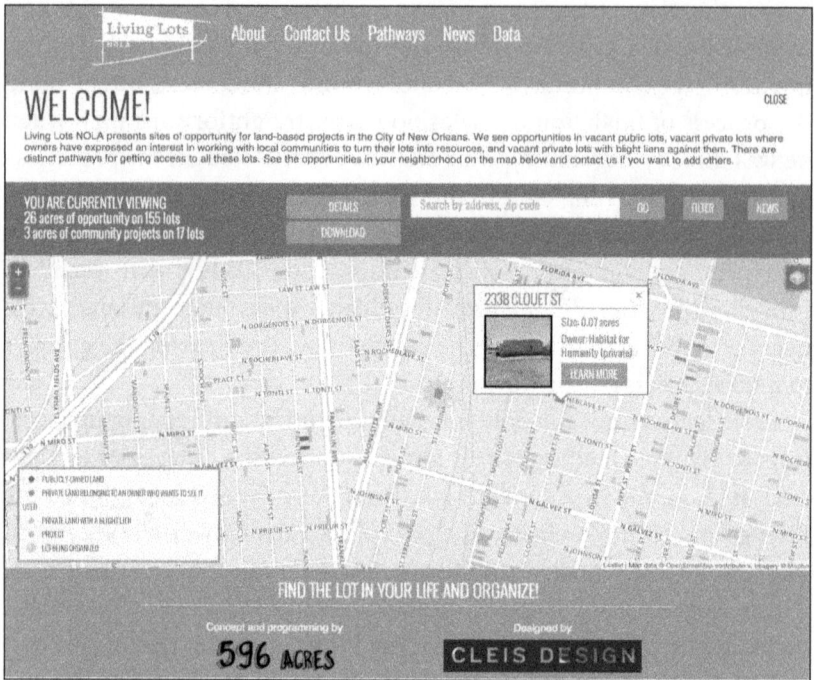

Figure 3.2. Screenshot of Living Lots NOLA website, captured by author on March 24, 2014.

land access even among those who were digitally capable. In the context of the rapidly transitioning urban context, growers seemed to prefer gaining direct information on their own, regardless of their socioeconomic, racial, or educational backgrounds.

Growers had reasons to be skeptical of the increasing external interest in urban cultivation and to be selective of the kind of land access they sought to establish. Grow Dat Youth Farm's successful establishment in City Park remained an exception even during the redevelopment period, with most growers finding it more efficient and reliable to identify smaller lots to lease or purchase from private owners and gradually build up their scale rather than starting with a single large space.

The failed collaboration between the Recirculating Farms Coalition and the Convention Center around 2012 exemplifies the risk of working with a large entity whose intentions and interests do not align with the growers'. The initial plan for the project was to establish an agricultural demonstration site on a five-acre plot owned by the Convention Center. The timing of the launch was to coincide with the first Farm to Table convention, which launched in August 2013. At the beginning of the plan, there were some interesting similarities to the Grow Dat Youth Farm's triangular agreement, with New Orleans Food and Farm Network and Tulane City Center partnering to assist in the initial development of the Recirculating Farms Coalition site. But after dragging out the negotiation process, the Convention Center eventually withdrew from the project, leaving the Coalition to search for another space.

The exact reasons why the Convention Center backed out remain unclear, but what is notable is that the Coalition's access to resources, including the staff's legal expertise, external funding, vetting of the ideas with various state and local agencies and the public, and the partnership with well-networked local nonprofit organizations and a university, did not safeguard them from having their access process stalled and failed. Recirculating Farms Coalition moved on from this project and eventually gained a lease to two lots within a few blocks of each other in Central City through a local development entity, where they would set up a demonstration garden that included an aquaponic system. The Convention Center lot was eventually developed in 2014 as a research site for LSU's AgCenter, though by 2016 the site was cleared and reverted to an empty, though maintained, green space.

Conclusion

The growers in post-Katrina New Orleans continued to face challenges throughout the decade, regardless of their aspirations, the timing of their entry into the scene, or their background and experience. Notably, the growers' struggle to find and secure land access persisted, from when the land was being undervalued and underutilized during the recovery period through the redevelopment period, when the state and private landowners began implementing land-lease programs. Their continued challenges resulted from trying to manifest something that does not fit the existing normative and legal categories of land use, food systems, or career pathways. In order to understand why the popularity of gardening or the availability of cheap, underutilized land alone did not lead to the emergence and persistence of urban cultivation, we must examine more closely how these opportunities, needs, and interests mapped onto the spatial and social landscape of the city.

Geographer Nathan McClintock theorizes the impact of urban agriculture by building on the Marxist concept of "metabolic rift," the idea that the capitalist appetite for natural resources will result in the alienation of humans from nature, as well as resource extraction that will cause irreversible damage to ecological systems.[15] McClintock argues that metabolic rift occurs not just at the macro-environmental level, but also in the form of failed social systems and individuals' sense of alienation under capitalism; thus, gardening appeals to urbanites when they feel a desire to reconnect with nature, food systems, or more physical forms of labor. In many ways, the destruction of the city during and after the 2005 storms was a physical manifestation of these rifts, which explains why urban cultivation "made sense" to a group of prospective growers. Yet this chapter illustrates that starting an urban cultivation project in the city where "it does not yet make sense" legally, economically, and socially, was no stroll in the garden, even for those who were otherwise in positions of privilege.

In their effort to manifest land uses that defied existing economic classifications of property value, growers looked to pragmatist solutions rather than political organizing when faced with these challenges. They prioritized starting their own projects over "fixing the system" through collective mobilization, and they exhibited a notable level of indifference toward the fact that they were practicing in legal gray areas or that there was significant variation among their methods and goals. They were willing to explore and experiment as they saw new opportunities arise, but these decisions were not entirely or solely opportunistic. Growers retained a healthy dose of skepticism about the existing system's ability to accommodate or adapt to their visions, as evidenced by their hesitancy to participate in government-run land lease programs and their preference instead to work with private landowners. These processes of adaptation and experimentation helped the growers assess their external and internal conditions, such as who was interested in their projects and for what reasons. Growers also were able to regularly reassess whether they still felt it worthwhile to pursue new visions of alternative urbanism. The next chapter further examines how the growers used urban cultivation as a prefigurative cultural repertoire as they continued to explore, experiment, and adapt their practice once they managed to launch a project.

4

Growing Changes in a Growing City

Forms and Characteristics of Prefigurative Actions

Nico was initially "planning on being here for ten days" when they arrived from Chicago in 2006 with a friend to drop off supplies and gather stories on the ground as a photojournalist. For a few years, before deciding to relocate to the region for the long term, they assisted with rebuilding efforts in Violet, St. Bernard Parish, about thirty minutes east of New Orleans. During this period, they contemplated extensively the significance of a young, white transplant such as themselves moving to a place where many people of color experienced severe displacement. But they finally said to themselves, "You know what? I love this place— why am I not just saying that I live here?" and took a job at New Orleans Food and Farm Network (NOFFN) in 2011 and found housing in New Orleans to formally relocate.

Through their work at NOFFN, as they helped build backyard gardens and set up community kitchens, Nico realized there was a vast potential and a need for building community-based programming centered around urban cultivation. Sometime in 2014, they began a search with their growing partner for a space to start a new cultivation project. During our initial interview in early 2015, Nico described their project in the Lower Ninth Ward, then in its early development, as follows:

> We found a lot, and it was in an area that we were interested in working in physically. And we also felt the need for local food in the Ninth Ward and the potential, like the fact that there's not really a grocery store, like a supermarket [there]. There's the food co-op, and there's Save-A-Lot. But we really felt like bringing fresh food would be really useful to the neighborhood. . . . My vision is, it comes from wanting to see neighborhood-centric growing projects.

But growing food was not the only, or even the primary, thing Nico wanted to do through their cultivation project. In fact, they added, "What I'm really focused on is composting." They saw urban cultivation as a tool to enact broader social reforms around food security and food systems. But they were not just working in urban cultivation. In order "to pay the bill," they also worked in construction and photography as they developed their ideas for alternative food systems, envisioning a composting network across the city, cooperative agricultural practices, and organically grown plant starters.

Nico began picking up compost from some restaurants to build their own composting project for the farm in the Lower Ninth Ward around 2014. The hoop house was set up on site, and they were getting it ready for growing. In late 2015, both Nico and their project partner got very busy with other jobs, and there was a "breakdown in communication." When the partner, who had more growing experience, left town, Nico realized that they needed to pivot. Coincidentally, at this time Nico learned that a staff member managing the compost operation at Hollygrove Market and Farm was also leaving the job and the city:

> It was all sort of serendipitous. She asked, "Does anyone want to take over this compost project? I pick up from a few restaurants and you get to trade for food and you can run a little education compost site." And I said, "Sure, this is the time when I am focusing on that more." And then I just went to all of the businesses [that she used to pick up scraps from and asked], "How much would you pay? [I'm] looking to create a business model from doing this." So I surveyed the scene and I started picking up more clients. I started out with five clients and that was three years ago, and now I have thirty-five clients.

Their composting business, Schmelly's Dirt Farm, has expanded its operation since its start, but Nico continues to prefigure new ways of operating as a small business:

> Now I am working pretty much full time. And I hired my first employee from Grow Dat [Youth Farm] and that was the goal when I was at NOFFN. There was the idea that all of these kids [are] going through

these high school agriculture programs, but then what jobs did they get with those skills if they are really passionate? . . . And we finally on-boarded [the recent Grow Dat Youth Farm graduate] and on his first day here, when we came over to the finished compost pile, he scooped up the compost and smelled it. And I was like, 'what seventeen-year-old does that these days?'

Nico's experience of unexpected trajectories, responding to serendipitous opportunities, and continuing to adapt and experiment were typical for many of the growers in this study. Their path reflects the pace of change that the growers, and the city, experienced over the decade, especially as the city shifted from the transitional period to the redevelopment period.

Making urban cultivation work in the city as a form of prefigurative urbanism was a challenge—but it was also a continuously evolving practice. From gaining lot access and setting up a growing site, to establishing a market for produce and learning how to grow at scale, all while facing the whims of the natural world: each required different skillsets and grit. Few people outside of the practice fully understand what it takes to sustain the work, even if they admire it. Many of the growers that I observed, like Nico, demonstrated their dedication to the work by showing up each day to tend the soil, even as they continuously faced decisions about where to take their urban cultivation project next. It was a never-ending process, partly because there was no "end point" toward which they were working in the end-effacing practice.

How do you prefigure an alternative way of life in the city through gardening and farming? This chapter illustrates the range of prefigurative practices that appeared in New Orleans's urban cultivation scene over the decade following the 2005 storms, with a focus on how growers expanded and experimented with their practice after launching their projects. Their decisions about whether to scale up or expand reflected their understanding of the changing opportunity structure and public interest in their practice, as well as their efforts to claim autonomy over a project's trajectory. Throughout, we see that growers mostly continued to approach their practice in prefigurative terms, focusing on their immediate goals of continuing to bring about direct change. They had no illusions about the difficulty of making a living as urban growers or about the limits of what their individual projects could accomplish. Toward the end of the decade, they became increasingly reflexive of the role they played in the city, especially around the race and class privileges that afforded some of them more opportunity and capacity to engage in urban cultivation as a form of prefigurative urbanism than others. But carry on they did. In a city where so many promises had been made with little action to follow, especially in the months and years following the disaster, at least they were doing *something, now*.

Redefining Urban Land Productivity

Presenting urban lots as places for food production at scale was a form of prefiguration by which growers demonstrated an alternative assessment of land's *productivity*. Land could do more than simply produce real estate values, or what Karl Marx defines as "exchange value," by providing a foundation for residential or commercial structures. Land could also have "use value," for gathering strangers to experience a sense of conviviality, offering a place of respite from urban congestion, or growing food to feed the community. The growers and the public did not always agree on how vacant or underused land might best be tapped for a productive cultivation practice, but the growers' practices tangibly demonstrated what they saw as the alternative "use value" of the land.

The most visible claim to this new form of urban land productivity came in the form of the physical transformation of a formerly abandoned or vacant space. As the previous chapter indicated, the initial process of preparing a lot for gardening or farming typically involved removing debris, installing fencing, and setting up infrastructure such as a water supply and tool sheds. Sometimes it included building on-site growing structures, from raised beds to aquaponic facilities, each of which visually contrasted with the surrounding area's residential or commercial buildings. By 2015, at least a third of the cultivation projects in New Orleans had a hoop house, a structure made out of PVC pipes covered with plastic sheets, though their size and number per site varied. In contrast to greenhouses, which are more permanent and costly, hoop houses are inexpensive, portable, and can be built in a day, especially with the help of a large number of volunteers.

Growers who elected to grow their crops in the ground had to undertake a much less visible transformation: soil preparation. One grower recalled the process of transforming her site in the Lower Ninth Ward as follows:

> So it took almost the summer honestly just to transform all of these lots into productive growing space. There was, like, a lot of trash that had to be cleaned up because people sort of had treated this dead-end as a place to dump trash. So we had to get all of that out and everything, and we had covered it with black plastic to kill all the grass and weeds underneath it. Black tarp had to sit there for six weeks to kill the grass underneath. And then [on] the other lot we built three beds. Once all of the grass and leaves had died, then we had a Bobcat come through and scrape the grass and any remaining soil. And then we brought in the actual soil and a bunch of coffee grounds, mixed that all into the soil, which worked better on some lots than others. A few of the lots had the house foundation underneath the soil, so that was pretty hard trying to get all of that stuff out of there.

Removing existing vegetation, remediating the soil of harmful toxins, and enriching it for cultivation prefigures a form of ecologically sustainable urban land management that prioritizes a lot's potential for growing food. The lot's value comes in the form of environmental sustainability and agricultural production rather than its potential for commercial development. Most growers could not afford to truck in sufficient topsoil to transform their lots, so they attempted to develop fertile soil using a combination of organic methods similar to what this grower did, including first killing the existing vegetation by solarization or occultation, planting sunflowers that were expected to take up lead from the soil, and then covering the space with woodchip or compost. This was a lengthy process that could take months or even years as was the case for Pamela's Sun Harvest Kitchen Garden described in Chapter Three. Moreover, it might not be clear to neighbors that anything was happening. In the meantime, soil remediation that takes place under the tarp is invisible, and a tarped lot could look like an eyesore or even blighted to passersby.

Growers continued to grapple with whether to prioritize productive cultivation over aesthetics even after they began cultivation. The growers who maintained a *community rebuilding* aspiration, for example, wanted to gain community approval of their work, which meant that they sometimes had to change their plans to meet neighbors' expectations, even if doing so would compromise the space's capacity to produce food at scale.

Zach practices permaculture at CRISP, which is a horticultural philosophy and method that aims to recreate natural ecosystems with diversity and resilience in agricultural practice. This means his space is not neatly compartmentalized in ways that Western European gardens tend to be, as the plants are situated to encourage symbiotic relationships to facilitate pollination, pest management, and improvement of the overall ecological system (Figure 4.1). But not all his neighbors shared his philosophy, and Zach took into account their wishes when deciding on what to plant:

It's not a bad thing when butterflies or caterpillars are coming and eating your stuff, because they're making butterflies and birds are defecating on your soil, and it all works together. And I think our problem as human beings is thinking we can control nature, thinking the nature is there to serve us, if you will, that causes the separation between us and the earth which causes separation between us and each other, which you see on Facebook and on the news, and everywhere. And that's my belief. . . . [But] I was talking to [a neighbor] across the street. She says, "That mulberry, that's just going to make things messy on the sidewalk." It brings birds and who cares about the sidewalk? But you have to respect where [they are coming from]. She doesn't want a mulberry tree, because she doesn't want a sidewalk like that. [Then I have to ask myself,] what might she want?

Other growers reported similar incidents when neighbors commented on "weeds" growing in a garden space (they were actually cover crops) or complained that a grower's cultivation project did not fit with the public's perception of a garden. The purpose of a cover crop is to restore the soil between growing seasons, or sometimes to

Figure 4.1. CRISP growing space in 2018. Photo by author.

complement the harvest through pest or moisture control. But to the casual observer, cover crops looked like overgrowth. David once received a note from a Habitat for Humanity representative stating that they would terminate the lease if he was not using the leased lot. He suspected that a staff member had simply driven by to check on the condition of the space and did not see past the tall grass at the front of the property. If she had stopped and looked more closely, she would have seen "tons of rows of crops among Johnson grass." He described this as an example of people having "different ideas about what a garden is," especially from the standpoint of varying practical needs and agricultural methods, including the choice to not weed extensively or to not prioritize the aesthetic appearance of the cultivation space for non-growers.

If the intended prefigurative impact of the physical transformation of the space was to demonstrate that a city lot can be used as a cultivation site with an intent to distribute, then one way to convince skeptics that it was possible to grow food at scale was to physically emulate a "farm." But this, too, could elicit opposition. Macon recalled that not everyone

Figure 4.2. Macon working at Hollygrove Market and Farm site around 2009. Photo by Amy Graham.

was pleased with how he grew rows of arugula and greens at Hollygrove Market and Farm (Figure 4.2):

> [The purpose of the site] was to teach market gardening on a small space. And in order to do that realistically, you have to sacrifice some diversity. You have to rotate your crop once a year, you know, you have to let the field rest in the summer, you have to come back with a different crop. You're not using a science of crop rotation, but you know if you are building compost [then] you're switching off. That's the trade-off. I got a lot of grief from people there that it wasn't more diverse. But the object was not to be sustainable but to make money, and you can't make money by half a row of this and a quarter row of that or twenty feet of—you know? You've gotta [sacrifice diversity].

Macon's production style stood out at a moment when most other growers were engaging in more biodiverse growing systems, whether driven by philosophy or curiosity. He compared his method to other growers, including the two mentor farmers that had since taken over his space, noting that diverse crop production required more labor. These other growers relied heavily on the labor of volunteers to maintain, harvest, and prepare various types of crops to sell in small quantities. His decision to sell two or three crops each season in bulk to restaurants made the process manageable for an individual, though he also had apprentices and volunteers working alongside him in the field. But the industrial agricultural aesthetics of Macon's space contrasted with the conventional notion of a "city garden," where one might expect to see a variety of crops and flowers arranged across the space, rather than rows of a single type of plant.

At the time of my study, the New Orleans cultivation scene included only a few sites engaged in controlled environment agriculture (CEA) practices, such as hydroponic or aquaponic systems. In 2012, Aquaponic Modular Production Systems (AMPS), founded by two Tulane graduates, attracted local media attention by designing a vertical aeroponics system. The company set up a demonstration system at Hollygrove Market and Farm around 2013; soon afterward, it installed another on the roof of the new Rouses supermarket that opened in the rapidly redeveloping Warehouse District, near downtown. Since its founding, the Recirculating Farms Coalition also has focused on an aquaponic system that incorporates fish into a water-based nutrient consumption and emission cycle. In 2015, the organization built a system at a Central City site with help from David, who already had a system at his own site in the Lower Ninth Ward. Our School at Blair Grocery also set up an aquaponic system on their original site around 2010. Despite the sustainability promises and unconventional appeal of water-based cultivation that would have been prefigurative, the costs and training required to build and maintain these systems did not appeal to the majority of the growers. Thus, they remained rare sights in the city well into 2015.

Composting was an alternative form of land productivity that some growers implemented on their cultivation site as a way to continue building soil and to manage green waste such as weeds, dead plants, and damaged fruit. The size and methods of composting varied across projects, and many cultivation projects had a small composting operation on site. The first large composting operation in the city was established by NOLA Green Roots, which began collecting compost from various businesses across the city during the transitional period. Nico's business, Schmelly's Dirt Farm, came onto the scene a few years later and rapidly grew its client base; city residents and businesses were beginning to embrace the idea of composting, but there was no citywide initiative for composting programs. Though composting produces soil and reduces food waste in landfills, this way of using city space does not fit conventional notions of property value. Furthermore, concerns over smell and rodent attraction raised questions about the adverse impacts on property values in dense residential areas. While operating in the legal gray area of waste management and urban agriculture, neither of which addressed composting explicitly at the time, compost operators like Nico and NOLA Green Roots manifested new land uses and sustainable practices before they were fully legalized or implemented in city policy.

Reimagining Agricultural Work

As a form of prefigurative urbanism, cultivation was for the growers not a means of agricultural production but a way of life that included the possibility of restructuring the relationship between work, labor, and compensation. This attempted restructuring involved exploring cooperative models and providing training and mentorship as entry points for those interested in pursuing the work full-time. There were significant disagreements over the merits and ethics of using volunteer labor across cultivation projects, revealing the varied interpretations of how to recognize and compensate agricultural work, especially in the context of the historical and persistent devaluation of agricultural labor in the US.

In keeping with their prefigurative approach, a few cultivation projects in post-Katrina New Orleans aimed specifically to model a new economic opportunity for the community through their aspiration to operate as a cooperative. Cooperatives are worker- or member-owned business that are managed democratically through a horizontal organizational structure. New Orleans has a rich history of cooperative-driven neighborhood developments that have mostly not been recognized by a broader public, especially by disaster response officials or outside cooperative activists, because they have operated mostly in working-class Black and immigrant communities.[1] Among the post-Katrina New Orleans urban cultivation projects, the VEGGI Farmers Cooperative in New Orleans East was distinct for operating formally as a cooperative. Daniel described its operation as follows:

> Everyone grows a combination of everything. So everyone in the co-op, their produce goes into some of the metropolitan markets, like for example, the restaurants, the Hollygrove [Market and Farm], the CSAs, Good Eggs. But also all the growers circulate [their produce] in this local community as well. So everyone has equal footing as everyone else, if that makes sense. So it's not just like, one grower, for example, is only gonna grow for Hollygrove and one grower's only gonna grow for VEGGI Co-op and one grower's only gonna grow [for themselves].

VEGGI's main retail clients were restaurants located in the wealthier parts of the city, twenty-five minutes from where the cultivation site was located. Putting the cooperative approach into practice required seasonal coordination on what to grow, as well as accounting skills to ensure that the revenue was fairly distributed.

A cooperative cultivation structure diffuses risk among the growers and allows them to sell collectively, in bulk, to larger clients. In the case of VEGGI, the managing staff members were mostly younger, second-generation Vietnamese or other Asian American immigrants who worked closely with older Vietnamese growers, functioning as cultural, social, and economic mediators in negotiating market agreements and land leases and in facilitating the cooperative management. Daniel described how VEGGI experimented with different payment structures for the growers as the cultivation project unfolded:

> That's also very tricky because we've been trying to figure out how we're gonna do this in terms of writing it down as a contract. In the beginning, it was more like a quota system like saying, "X percentage of your land has to be dedicated to the production for x purposes, x percentage has to be dedicated to x purposes." But now it's kind of—we decide on what's gonna be grown. We take what's grown in total and we split that, if that makes sense, between different venues. And we pay out according to the portion of what's been grown for each grower. So [if grower

A], for example, grows half as much as grower B, then they're gonna be paid in quarters to that proportion. So you could make a $100 for example, grower B is gonna make $75, and grower A, $25.

The cooperative model presented the potential for urban cultivation projects to scale up, generating significant economic opportunities within the local food system. Notably, however, Hollygrove Market and Farm's aggregation of produce and other agricultural products from rural and urban small-scale farmers, for sale to urban consumers, did not lead to the formation of a cooperative across these producers. Growers' lack of interest in forming a citywide or regional cooperative reflected the diverse aspirations and forms of cultivation across the projects and growers' general preference to retain independence. The lack of designated leadership also posed challenges in terms of coordinating across dozens of cultivation projects.

The growers operating with the *urban cultivation expansion* aspiration saw sharing their knowledge of cultivation practices, including the economics of cultivation, as part of their prefigurative practice. Many of the growers we interviewed frequently named Macon, Pamela, and Jeanette as their mentors. I began hearing about people "apprenticing" or "interning" at urban cultivation sites around 2012. When we spoke with Caroline in 2014, she was about to welcome two new interns to Grow Me Somethin':

> I guess about a month ago, I have one person that could come with me. I'd say working . . . Apprenticing? I guess? Interning may be the best way to put it. A friend of a friend who kept saying, 'I wanna help on the farm, I wanna help on the farm,' and I told her 'I can't pay you,' you know? I'm not even paying myself. And she came back to me and said, 'I don't care. I wanna know what you know. I wanna learn how to grow.' So she comes maybe for four to five hours a week. And last week she showed up with a friend of hers. So hopefully next week he'll be able to start. It's nice. It's nice that people are interested enough to wanna come spend time with me. I'm very flattered by it.

Even though *urban cultivation expansion* was not her primary aspiration, Caroline appreciated that she was attracting others' interest while getting free labor. But the fact that these "growers in training" were able to work without pay, even for few hours a week, suggests that these "opportunities" were not equally accessible to everyone interested in learning to cultivate food in the city. Sociologist Richard Ocejo's study of mostly white, male, middle-class young adults who redefined working-class jobs as 'artisanal' craftsmanship found that many of these individuals had to be able to live on little or no pay as they developed their skills in their chosen industry.[2] This type of flexibility required financial and social stability, even if the skills they were learning might eventually pay off in the high-end consumer market. The pathways for interns and apprentices in urban cultivation were less clear, because—unlike the butchers, mixologists, or hair stylists that Ocejo studied—there was no existing specialized industry where growers could cash in their new skillsets. While the racial and gender demographics of these interns were relatively diverse, they were predominantly younger transplants who had other sources of income, including familial support, to be able to do the work without getting paid.

Most urban cultivation projects relied on volunteer labor for regular site maintenance or irregular extensive work, whether erecting a hoop house or clearing the lot. During the recovery and transitional periods, these volunteers mostly came from the immediate neighborhood or had previous relationships with the growers. Toward the end of the decade, however, many of these volunteers were visitors to the city who came either as part of a leisure or professional trip to New Orleans, or specifically to work on an urban cultivation project. The growers who worked with these one-time volunteers listed the numbers and size of the volunteer groups that helped them on site throughout the year on their websites and reports. Many of the volunteer visits took place when the city's tourist season coincided with the cultivation season, including college spring breaks and the Jazz and Heritage Festival from March through early May. For example, Jeanette had a group of volunteers from HandsOn help her with preparing a new space she leased from Habitat for Humanity around 2013. They "cleared the property, covered it in six-mil plastic and mulch, started building the rainwater collection and the raised beds." Jeanette also had developed relationships with a few out-of-state universities, including the Ohio State University and Marquette University, whose students would come work with her several times a year, in groups as large as seventy-five for each trip, either as part of an extracurricular activity or to earn service-learning credits. Most of these visiting volunteers had little experience in cultivation, though occasionally one appeared with much-needed skills and resources, such as legal expertise or truckloads of mulch or tools. To accommodate this sporadic availability of labor, the growers would set aside types of work that would benefit most from a single day's work by a large number of mostly inexperienced volunteers, such as clearing the space, covering the lot with plastic or mulch, or building hoop houses.

Our School at Blair Grocery took it a step further by engaging in "voluntourism." Groups of students would come from New York City for more than a week at a time to work and learn while staying at the project's main building. These volunteers were expected to pay fees for their experience, which included room and board along with some field trips and discussions led by the staff. Erin, who left the project in 2011, recalled how volunteers became one of the key sources of income for the project:

> Volunteers would pay to volunteer or would have to make some kind of donation. And when we had the students from New York City, those volunteers in particular would pay a couple hundred dollars per individual just to pay toward the space. That's my understanding. So it was about being able to facilitate an experience, but it often was more time, we'll have to find things for them to do, as opposed to doing things that needed to get done. But we would turn a huge amount of compost with volunteers, which was great. It didn't need to happen, but it happened, that was great. Or we could develop a vacant piece of property in a very short period of time, by hand, which was cheaper than renting a Bobcat. But [it] could have happened with a Bobcat with a single person. It doesn't really take a lot of people to manage a farm space.

As Erin pointed out, charging the volunteers for the "experience" meant the project had to prioritize the satisfaction of these helpers by ensuring that they felt they were doing *something* good for others. These volunteers were participating in what anthropologist Vincanne Adams calls the "affect economy," which capitalizes on the empathetic reaction of the privileged class toward the less fortunate, often produc-ing more emotional satisfaction for the participants than benefits for those on the receiving end of the goodwill.[3] Prefigurative urbanism's tangible and immediate transformative promises makes it particu-larly attractive to those who wish to engage in the affect economy, and cultivation projects provided these experiences by sometimes setting aside or coordinating work that would provide suitable experiences for these visitors. Some growers candidly said that hosting these volunteer groups, including local university students seeking "service learning" credits, became its own work that had to be coordinated and man-aged. It was not always worth the income, publicity, or organizational networking that it yielded.

Other growers rejected using volunteers on principle: laborers should be paid for their work. Jamal explained why he no longer had volunteers for his project:

> We don't do volunteers. We've had volunteers. It's just . . . it doesn't work out. The paid work goes a long way especially in a working-class neighborhood, like why, on that front, why fuck with it? Because they're working.

From his perspective, having volunteers on site disrespected the people who were working hard to make ends meet in the immediate neigh-borhood in the Lower Ninth Ward. Having people who can afford to volunteer on site disconnected the project from the community, and expecting people in the community to work for free in their spare time was unreasonable. Others, like Macon, noted, "the success of the gar-den doesn't depend on having volunteers. I can manage this space." For Macon, the point of having volunteers was to provide a space for con-necting with other like-minded people, not for supplementing labor. Other growers critiqued peers who used volunteers for growing produce to be sold in the market.

The sector's reliance on volunteer labor highlights the challenges growers faced in making a living from their cultivation practices. Eric, who had an agreement with Café Hope's restaurant that allowed him to park his RV and eat at the restaurant whenever he wished, responded as follows when asked to describe his social class standing:

We had a meeting not too long ago with the two main growers, myself, the executive director, the chef, and maybe a couple of other people, and I basically drew out what I was gonna do. And nobody disputed it, nobody said anything. And I just kind of put myself on an honor system to only eat [at the restaurant] when I'm working in the garden or if I do the computer work, because I do their computer work also.

Eric said he ate "like a king" because of his access to his own produce and various restaurants in the city where he could bring produce to barter for a meal. Other growers also shared that they sometimes got free meals from the restaurants to whom they sold, including some fine-dining establishments that they could not otherwise afford. For these growers, cultivation afforded them a uniquely privileged lifestyle that defied conventional connotations of class and status. These privileges were ephemeral in contrast to their general financial insecurity due to limited income generated by their cultivation projects.

The growers' gravitation toward the term "growers" reflects their wish to distance themselves from the cultural and economic connotations of the terms "farmers" and "gardeners." They often made a point that they were not "gardeners," because gardening is associated with personal leisure, and they saw themselves as growing at a larger scale and with the intention of feeding others. Yet these growers also hesitated to call themselves "farmers," because they did not feel it appropriate to equate their small-scale operation in the city to large-scale agricultural production farms in rural areas. This hesitancy came out of respect for large-scale, rural farming as an occupation; it was not an attempt to dissociate themselves from the social devaluation and the economic struggle of many farmers in the US. Some of them specifically wanted to distance themselves from the term "urban farmer" itself; one grower explained why:

I don't know. It might be the personal imposter syndrome stuff. It's implied, sort of, I don't know, there's a pretentiousness to it. That just might be perceived on my part as well, but specifically within the urban farming community, there's whole lot of idealism there with not a lot of actions. . . . But I like being boots-on-the-ground and hanging out in the dirt and all of that. I feel like the term's been weirdly subverted, but again I don't have another word for it, so.

Similarly, an otherwise simple demographic question about the growers' class identification often prompted long-winded self-assessments as Eric's response above exemplifies. The growers typically conveyed that they were not financially well-off, but that their work provided them with a "rich" life through access to high-quality food and physical and mental health through working outdoors. Some considered their practice a reimagining of agricultural work. Colleen, who was enrolled in a master's program at the time, lamented the devaluation of agricultural work:

> No, it's not valued. As soon as I'm not a student anymore, I'm going to have to get a big-people job, because I certainly am not going to make enough money on this space, or I have to put the energy, effort into another space to make it profitable. But that's just getting unpaid. But I've learned to live poor. It's great. . . . I have the best food ever, you know? I mean, really, what do you want in life? You want some good food and some good friends. And for me, I want to spend some time outside. I have that. I mean, I'm not going to buy a house off that. I don't have any kids, so I'm not going to put them away to college on that. But it's a nice life, really.

The growers most likely to make these types of statements often had some kind of financial safety net, whether in the form of family support or alternative sources of income, underscoring their capacity to have a "choice" to pursue an alternative lifestyle and career. These growers recognized their privileges at the same time that they emphasized that cultivation did not by itself provide sufficient economic means to sustain their lifestyle. By contrast, the growers who lacked financial security approached urban cultivation with more pragmatism; they were less likely to put a positive spin on their social class standing. Instead, they tended to describe their financial insecurity or challenges in matter-of-fact terms, even when asked about the long-term prospects for their cultivation projects.

Cultivating New Markets

As a form of prefigurative urbanism, growers' cultivation practices continued to change in scale and scope as they adapted to changes in external conditions or internal interests and capacities. This tendency toward experimentation was most visible in how they engaged with commercial production. Even growers who had not originally planned to sell their produce began exploring the possibility when new economic opportunities emerged. At first, their efforts to create new markets was slow going. But over the course of the decade, as growers experimented or adopted new techniques, plants, products, and market venues, their focus on selling expanded.

The opening of Hollygrove Market and Farm in 2008 signaled new interest in and support for locally grown food as the city headed into the transitional period. Prior to this, there had been no established distribution system for produce grown in the city, and the handful of growers who had started their cultivation projects during the recovery period were mostly giving the food away to nearby residents or consuming it themselves. Yet, when presented with the opportunity, only a few urban farmers sold to Hollygrove Market and Farm on a regular basis. Likewise, urban growers rarely sold through Crescent City Farmers Markets, which had been in operation since 1995 and quickly resumed operations in November 2005. The growers told us that the farmers markets worked better for rural farmers than they did for urban growers, because the market organizers expected year-round, consistent production.[4] Growers also regarded the expected volume of sales as an obstacle to participation, especially for the types of crop that require a lot of land and a long term between planting and harvesting, like squash and melons. For these and other reasons, growers were more likely to develop ways of marketing their products on their own, typically to restaurants or to individual consumers.

During the transitional period, those hoping to sell their produce began experimenting with farm stands or CSA programs. Ashley recalled how she distributed the produce and eggs she was harvesting at Trouser House around 2010:

Everything was marketed at the monthly events at the farm on second Saturdays. So people knew. People would come out and see. [They'd say,] "Oh, what's growing?," and "How is it doing?," and, you know, it was kind of fun when someone would come in August, and then they came back two months later and they were like, "Oh my gosh, that basil is like as big as my shoulder. That's crazy!"

The aim of these stands and small markets was not necessarily to sell at volume, but to bring people out to the space and showcase what the project had been growing. Those who operated farm stands did not necessarily keep regular hours, and they tended to announce their presence through homemade signs, word-of-mouth, or online announcements. The CSAs mostly had only a small number of subscribers and were established through personal connections or references. The exception at the time was Grow Dat Youth Farm, which operated a CSA program with citywide public participation.

Increasingly during the redevelopment period, growers with entrepreneurial aspirations (*alternative career* and *social entrepreneurialism*) began wholesaling their produce to various restaurants in the city. These restaurants were primarily located in quickly recovered or redeveloping areas, such as the French Quarter, the Warehouse District, or Uptown. Most were new restaurants that had opened after 2012, but some growers located clients among a few long-standing fine-dining establishments. The process was exploratory and experimental, as each grower developed their own relationship with chefs. The growers discovered that growing specialty crops or highly perishable items for a chef or a restaurant group was a niche market that would not compete with the lower cost and high volume of existing distribution channels.

Without an existing network of restaurants buying from urban growers, finding clients and negotiating the terms of sale was a time-consuming process. The growers had to learn to market and ascribe value to their products. Taking cues from the popularity of the "Farm to Table" concept across the US during this time, these restaurants proudly noted local ingredients on the menu. Yet it was the growers' persistent marketing to individual chefs that kept the transactions going, even shifting the power dynamics of the producer-supplier relationship. As chefs increasingly requested specialty items, there was always a risk of the growers being left with the produce that restaurants passed on. Jamal described the type of relationship necessary to sustain the "farm to table" system:

> There's like a new generation of restaurateurs, a lot younger that have that social aspect to it, and they want to support the farm to table concept. [Names a few restaurants.] [The chefs would ask,] "How come you don't

sell more stuff to us?" We used to use restaurants as the last option or a dumping ground. We have too much kale or too much radishes, [then we would ask the restaurants,] "Do you want to buy some?" They're like, "We'd rather buy a lot. What do we need to do to buy more?" They'd start buying seeds, it's like locking it in. The farm to table concept, too. I don't know what people think about it, but if the chef is not involved with the farm, and if the chef is not changing the menu based on what the farm is doing or other farmers, then I don't really see it as farm to table. It doesn't have to be one hundred percent all from local farms. One guy, he wants to pay for a micro green setup for that at my house. If you're going to buy it, sure, and pay for it, why not, I'll do it.

Growers had to find ways to reorient the seller-buyer relationship with the restaurant to share the cost and risk of the production. Most chefs and restaurateurs were not easily persuaded to make this kind of commitment. The growers who insisted on these new terms managed to find a few that were willing to work with them, but even these arrangements tended to take the form of handshake agreements rather than written contracts, thus presenting ongoing risks for the growers.

Growers who elected to sell their produce found themselves operating in a legal gray area, at least when it came to city code. Before the 2015 revision of the comprehensive zoning ordinance that included a distinct section on agriculture, the only legal form of commercial cultivation was "farming," defined as:

> Farming, including the usual farm buildings and structures, and animal raising, trapping and fishing, on sites of five (5) acres or more provided such use is not in conflict with any other ordinances of the City of New Orleans.[5]

The zoning ordinance did include provisions for smaller cultivation projects, but only for "propagation and cultivation," and not for sale on site:

> Private gardens, truck gardens, and nurseries for the propagation and cultivation of plants, only when said plants, flowers, or produce are not offered for sale on the premises.[6]

Such language definitively precluded on-site farmstands, but it left room for interpretation when it came to selling directly to restaurants. Legal uncertainty, in other words, did not stop growers from engaging in commercial production.

During the redevelopment period, growers increasingly expanded their commercial profile into non-produce items, including animal husbandry, beekeeping, flower growing, and composting. Two cultivation projects that exclusively specialized in flowers both started in 2014, selling flowers directly to consumers and to events and venues, in addition to a flower CSA. David turned to beekeeping and cultivated new services and products that became additional sources of income for Capstone through honey production and hive removal. His interest in honeybees came out of his observation that there were very few honeybees in the area where he was growing. Someone gave him his first hive, and then he secured funding to obtain more. When asked how he learned to tend honeybees during our follow-up interview, he responded that it was "self-taught" with a "lot of research." He continued:

> My goal for this year was to have twenty hives because I realized last year, when we had ten, that honey was a great source of revenue, and so I wanted to double that and expand our revenue. My goal was twenty hives, and we're currently at twenty-six hives. Well I bought several when I won Pitch NOLA Living. Well, actually I came in second place, but I bought a lot of equipment with that money, and I've also started doing beehive removals from houses and swarms and if it's somebody here in the Lower Ninth Ward, I typically don't charge any more than what my expenses are.

He quickly became the go-to bee remover and apiary mentor in the city, especially after the *Times-Picayune* published an article about him and Capstone in 2014.

When Good Eggs opened one of its three new branches in New Orleans in 2013, few growers immediately signed on to sell their products through its online market. The San Francisco–based online farmers market offered web-based sales that made it easy for customers to order small quantities of locally grown and produced food, plants, and value-added goods, with an option of having the items delivered. Rather than competing with this new platform, Hollygrove Market and Farm partnered with Good Eggs to outsource some of the delivery services that it had been operating in-house. Growers appreciated the possibility of gaining new customers through the platform's digital infrastructure, but many still had reservations about signing up. Jordan took issue with the company's sales fee, which he noted was thirty percent of the sales, among other concerns:

> If somebody orders [and says], "I want a stalk of sugar cane and a bunch of rosemary." Then you'd have to drive all the way out to [one of the cultivation sites], 'cause that's where we were doing the sugar cane. Cut one stalk, and drive all the way to the Urban Farmstead site to get rosemaries. Cut a bunch of rosemaries. Then you'd have to drive to Mid-City and give them a bunch of rosemary and one stalk of sugar cane.

While some growers like Caroline and Jeanette were willing to take a chance on this new platform and for a while had a relatively positive experience selling through the website, many others did not find this distribution channel worth their time, especially if they had already set up their own distribution to restaurants or individual consumers. Their skepticism was validated when the business abruptly shut its operations in New Orleans and two other cities in 2015, after less than two years of operation.

Nearly all nonprofit cultivation projects eventually participated in some form of market activity. They did not view this as contradictory to their nonprofit mission, as they found ways to make sense of commercial production as an income stream for fiscal independence, an educational opportunity for youth, or a valuable experience for adults interested in commercial production. The growers' varying and mostly individualized approaches to creating new distribution systems for their products emerged from their realization that the existing market did not accommodate them or value their products appropriately. Their foray into commercial production neither rejected capitalism nor embraced the status quo. But their experimental approach prefigured alternative systems of food production and distribution in the city by redefining the value of their labor and products. They prefigured what needed to exist without providing ideological justifications for their practice, and they continued to improvise or modify as they learned what did or did not work so they could keep selling what they grew.

What Do You Grow, How, and for Whom?

As growers' awareness of their shifting capacities and their role in the city changed, so too did their decisions about what to grow. The growers who were new to cultivation learned from their mistakes, while those experimenting with commercial sales had to figure out what crops best suited restaurateurs' tastes. Every season presented an opportunity to try something new, from plant selection and seeding timing to climate adjustment and pest mitigation. The growers disagreed over what they considered to be the *best* methods, but there was a general expectation of and respect for diversity in practice.

The growers engaging in commercial sales quickly learned that they needed to grow what sold well, by trial and error. Caroline recalled what she described humorously as "the great basil lesson of 2014," when she realized that the four kinds of basil she was growing and selling well in the spring became less profitable later in the season, when other growers began selling the herb at the same moment that consumers themselves were growing it at home. She added:

> It's just sad, 'cause they're the most beautiful basil plants I'd ever grown in my life. But it taught me a good lesson in product planning and understanding what's already been produced locally, what other farmers are growing. And really focusing on growing unusual varieties, because that's what I had in the beginning at Good Eggs. I was growing calendula flowers. Nobody else had calendula flowers. Now everybody on the market grows basil. . . . So all the stuff that I have besides basil are those heirloom varieties, the unusual shapes, unusual colors, so I think I will do much better, the sort of luck I was having in May as far as moving the stuff. So, you know, lessons learned!

Commercial growers like Caroline adjusted their crop selection to meet the needs of the market, often focusing on specialty plants and herbs that chefs or other customers expressed interest in buying locally.

It made economic sense to not compete with the much cheaper flows of produce arriving from global or regional supply chains, even when growing specialty items came with risks of its own.

For the growers who began their cultivation project with the *community rebuilding* aspiration, the priority was to grow food that long-term residents, especially working-class Black New Orleanians, wanted to eat. In the late summer of 2014, Jenga described what was growing at the two gardening sites she helped start:

> Cabbage, collards, broccoli, different things that are kind of appropriate for fall, you know. We'll give those away. But outside of that, we really don't dictate what anybody grows. We have a community plot at the Guerrilla Garden that we kind of maintain. Every season we grow something different. Usually in the fall, we grow greens, and that's to give away. You know, it's a really good way to kind of bring people into the garden. People love free stuff, obviously. And then it helps to encourage people, too, to eat fresh, local, organic food, you know, that they can get for free at the garden. So like I said, we're really not like a production farm. We don't have production goals of growing *x* amount of produce. We're really more on the social and community-building side of it.

Aside from growing what appealed to buyers' or neighbors' tastes, all growers, to varying degrees, expressed curiosity about trying out certain species of crops. Colleen sold some of her produce to Hollygrove Market and Farm. She described her decision-making process on what to grow as follows:

> I love certain crops. I'm obsessed with them. The way I grow my lettuce is beautiful, and I love learning a new plant, and then I want to get really good at that plant. Then I have the nematode (plant parasite) issue. So, right now I'm getting ready to manage that. . . . There are certain sections of the gardens that I grow just for volunteers. So the blackberries are for volunteers. The muscadines, I sell some of them, [and] we made wine this year [out of muscadines]. I gave away the wine and some of it is just, again, I'm not trying to be a farmer, but also if you get people excited, and you send them home with food, that's the value, you know? I'd rather give it away than get paid two dollars a pound for it and hustle.

These growers saw urban cultivation as an opportunity for their own personal skill development or a way to educate the public on the joy and potential of horticultural practices, thus they were not willing to only grow what was most profitable. As Caroline's ability to market specialized crops exemplifies, the emerging new food economy's interests in heirloom or non-European plants often aligned with the growers' own interests in exploring what was new to them. Notably, only a couple of growers mentioned reviving native plants as an aspect of their exploration, often due to the expected benefits for water and fertilization efficiency.

For the growers who attempted to engage in both commercial production and community building, balancing multiple and sometimes contradictory tastes became a challenge. VEGGI Co-op was an exception in this case, as the Asian produce its Vietnamese growers harvested appealed both to its core buyers in the local Vietnamese immigrant enclave as well as "metropolitan markets" like restaurants, Hollygrove Market and Farm, and individual CSAs. But in most other cases, growers had to prioritize growing some crops over others, given their limited cultivation space. The produce that long-term residents, especially Black New Orlenians, preferred, such as collard greens, peppers, and sweet potatoes, took up significant space, time, and energy, while high profit-margin produce like microgreens and herbs grew quickly in small spaces throughout the season.

Coincidentally, restaurants were more willing to buy microgreens and herbs because they are highly perishable, and are therefore best sourced locally. In anticipation of criticism that selling microgreens to fine-dining restaurants did not align with the organization's frequent references to food justice, Rob justified Our School at Blair Grocery's approach as the "contemporary Robin Hood approach":

> 'Cause our neighbors are not going to eat arugula or radish sprouts. 'Cause they look at them and they're like "What?" They want to eat collard greens, mustards, tomatoes, eggplant. Regional southern things. And we can't really grow those. Well we can, and we're getting there. Now that we've expanded. But we grow enough of it so that our neighbors can come by anytime and get something that they want. And the way that we deal with the cost of that, 'cause it takes our time and money to grow

the stuff. And it's like a losing battle, a revenue-losing prospect if you're just giving stuff away all the time. So we sell to the restaurants uptown that we can charge a good amount of money for stuff. . . . And so when a neighbor comes by and wants something, like say they want a bag of collard greens, you know we'll bag up two, three pounds or whatever and they'll say "How much?" and we'll say "I don't know, what do you pay at Walmart?"

While not exactly naming it as explicitly as Rob did ("Robin Hooding"), the growers who sought to rebuild community often justified their sales to fine-dining establishments or wealthier consumers as a way to subsidize the lower prices of produce for the neighbors. The conundrum over prioritizing what to grow in finite space reflected the divergent definitions of "local food" by long-term Black residents and white newcomers with alternative food movement interests.[7]

In other cases, growers' changing plant selection stemmed from their changing capacity. Zach explained why he was planning to focus on low-maintenance plants on his newly acquired lot:

It's changed from what I originally thought, which [was], 'Okay, we'll try and grow vegetables and kale and collard greens and sell them to restaurants and sell them to different people and have people come work for us.' [It has changed] to more as where I want to find plants that do well without much care at all, maybe some of them come from here, some of them don't, so we have a malabar spinach that comes back every year, scarlet runner beans, mulberries, obviously. Different fruit trees and things like figs that don't need much care. And we've got these chickens that I just grabbed six eggs from yesterday, so I want to do more of that and make kind of an edible forest.

It had taken Zach a couple of years to formalize the ownership of this new space, and by that time he had taken on another job that reduced the amount of time he could be in the garden. His new priority was to focus on low-maintenance, long-term, and high-yielding plants like fruit trees and perennial plants that do not require constant tending.

Growers' practices varied not just in terms of what they grew, but how they grew. Most cultivation projects used raised beds, which were constructed out of rocks, cinderblocks, or two-by-fours, both to isolate the growing soil from ground contaminants and to create better drainage. But gardeners disagreed over whether raised beds were necessary if soil testing proved the ground appropriate for cultivation, as well as over the optimal time to start planting, how closely to space plants, and how best to mitigate pests and other diseases. Younger and more novice growers referenced workshops and online resources such as YouTube tutorials, whereas older and more experienced growers relied on their preferred method for raising crops in the subtropical climate of southeastern Louisiana. Jeanette lamented that many new growers in the city were not learning techniques developed by long-term growers like herself and Pamela Broom:

> The people that I have interacted with on the subject [of gardening], I am disappointed with the follow-up, because there are so many things that could be done. For example, I would like to see, I wish I could have seen more people looking at the Garden on Mars when it was completed. Because the idea was so incredibly simple. And the value of it was just amazing. The idea, I mean, so much thought and discussion have been put into that project to make it, it is near perfect as we possibly could [make it]. And we couldn't get people's attention. . . . [Growers with knowledge and experience] are just too few. And what we have out there in the greater numbers are gardeners who are not using good techniques. My friend Pam Broom is a gardener who uses very good gardening techniques. And I don't understand why the message isn't getting out there.

Of the growers interviewed for this study, only eight had bachelor's or advanced degrees related specifically to cultivation, such as horticultural science or environmental law.

The growers seemed open to sharing their knowledge with anyone curious enough to listen, rather than guarding their secrets, and I heard of their giving advice or material support to each other on occasion. Each grower had a strong opinion about what to grow and how, and a diversity in practice was expected without any cultivation project becoming the hailed model for all others to follow. The increasing trend of seeking solutions in national or even global sources, both based on academic research and tried-and-tested methods, further disconnected post-Katrina

New Orleans urban cultivation from its historical precedent, as Jeanette observed. On one hand this reflected the broader demographic shift in the practice, with the increasing dominance of newcomers working in urban cultivation. But climate change required all growers to experiment and adapt as extreme weather began to pose imminent threats to their practice, elevating their urgency to innovate and experiment by looking for answers elsewhere.

Engaging the Public

Beyond converting the space physically and selling or donating their produce, growers tried to engage with the public to share how gardens and farms fit into their visions of alternative urbanism. Their formal and informal efforts to engage the public differs from social movements' approach to organizing, as the growers were typically not focused on mobilizing a constituency around a specific collective cause. Rather, they explored a variety of forms of engagement to various ends, from hosting free spaces for social interaction to offering paid workshops on horticultural techniques. Some of their efforts successfully engaged the segment of the public they sought to bring to the cultivation space; in other cases, the extent to which they could "make a difference" in the lives of these individuals remained unclear. The growers' engagement with the public reminds us that while prefigurative urbanism is conceived of and executed individually, its adherents intend to bring about some form of collective or public change. In part because of the open-ended, exploratory nature of the practice, the extent and the nature of these changes are not dictated entirely by the actors themselves.

One of the most common ways that urban cultivation projects tried to invite the public to show up and start engaging was through public social events. The intentions and the format of these events varied depending on who the growers were trying to attract to the cultivation space. For the growers aspiring to rebuild community, social events brought in long-term residents to see the cultivation project and to spend time with each other. Reaching these residents, many of them senior citizens who did not use electronic communication or had limited mobility, required time and energy. The growers would go knock on doors and hang physical flyers to inform nearby residents about upcoming events at one of the two gardens. Jenga explained how her organization approaches community outreach:

> Our project is really focused on the community aspect of it, so that piece is very, very, very important to us. So we communicate with our neighbors through, like, door-to-door outreach. Like I said, all of our programs, all of our activities in the afternoon are free, and open to the public. So that's another way we communicate with people. We have an email newsletter that goes out. We focus on Lower Ninth Ward residents and people who have come to workshops and things in the past, because we really want to be kind of a gathering place for people in our community. There really aren't that many places. I mean, that's not true. There are a ton of churches in the Lower Nine. But really, outside of churches, there aren't many gathering places, or community centers, and we really see the Guerrilla Garden as a community center.

Jenga's comments underscore that this type of urban cultivation was intended to prefigure a new social space for the community, whether or not community members were involved in gardening. So, the growers needed to proactively invite those who might not otherwise seek out an opportunity to engage.

Social gatherings could become a starting point for a larger conversation about issues facing the community,[8] though most social events focused solely on bringing people together, with or without a thematic focus. Similarly, seasonal events and weekend or summer programming for young people were intended to bring community members out to witness the gardens' work and to gain their support for the projects. Workshops on gardening, nutrition, and cooking more often attracted participants from outside the immediate community, especially when most of the promotion relied on online communication, including email newsletters or Facebook posts.

Cultivation sites also became social event spaces for commercial or fundraising events, making them spaces of consumption as well as production. Paradigm Garden regularly hosted dining events or outdoor yoga sessions. The yoga classes incorporated their two farm goats as assistants in "goat yoga," generating new income streams. Cultivation

projects frequently partnered with local artists and venders who provided music, food, or drinks for the events. Revelers would be given
a tour of the gardens and were encouraged to take photos in front of
the flora and fauna. Growers of these cultivation projects prioritized the
space's aesthetic appeal and had to leave sufficient space to set up tables
and chairs for events.

Nonprofit cultivation projects often held fundraising events at their
own cultivation spaces, recognizing the appeal of the "party in the garden" aesthetic for white, middle-class urban consumers, especially those
following the alternative food movement.[9] The tickets for these events
ranged from $30 to over $50 per participant (Figure 4.3 shows a poster
from a fundraising event at Hollygrove Market and Farm), a price that
put these events out of reach of the majority of long-term residents of
New Orleans, where the median household income in 2015 was less than
$40,000.[10] These events tended to emphasize the aesthetics of the white
"agrarian imaginary," with string lights and cut flowers in Mason jars,
and were free of any reminders of the labor and land exploitation that
has long been at the root of the agricultural industry in the US.[11]

Many cultivation projects' public engagement events focused on
young people. Programming for younger children tended to focus narrowly on either exposing them to nature or teaching healthy eating habits. After the charter school Edible Schoolyard was established in 2006,
other public and private schools across the city also set up gardens on
school grounds. These schools hired growers part-time to maintain and
run educational programming on site, and both for- and nonprofit cultivation projects hosted kid-centered events such as school field trips and
summer camps. Only a few of these garden beds lasted for multiple academic years, because they relied on a dedicated teacher or administrator
to maintain the efforts. Without the school or the city creating dedicated
resources for these gardens, whenever funds ran out or the teacher or
administrator left the school, the garden often became overgrown or left
empty.

Despite the popularity of bringing young people into the gardens, it
has not been easy to measure the real impacts of urban cultivation programming on children. Images of smiling Black children holding up car-
rots or tomatoes that they had presumably just harvested in the garden
grace the websites and brochures of these programs, signaling hope for

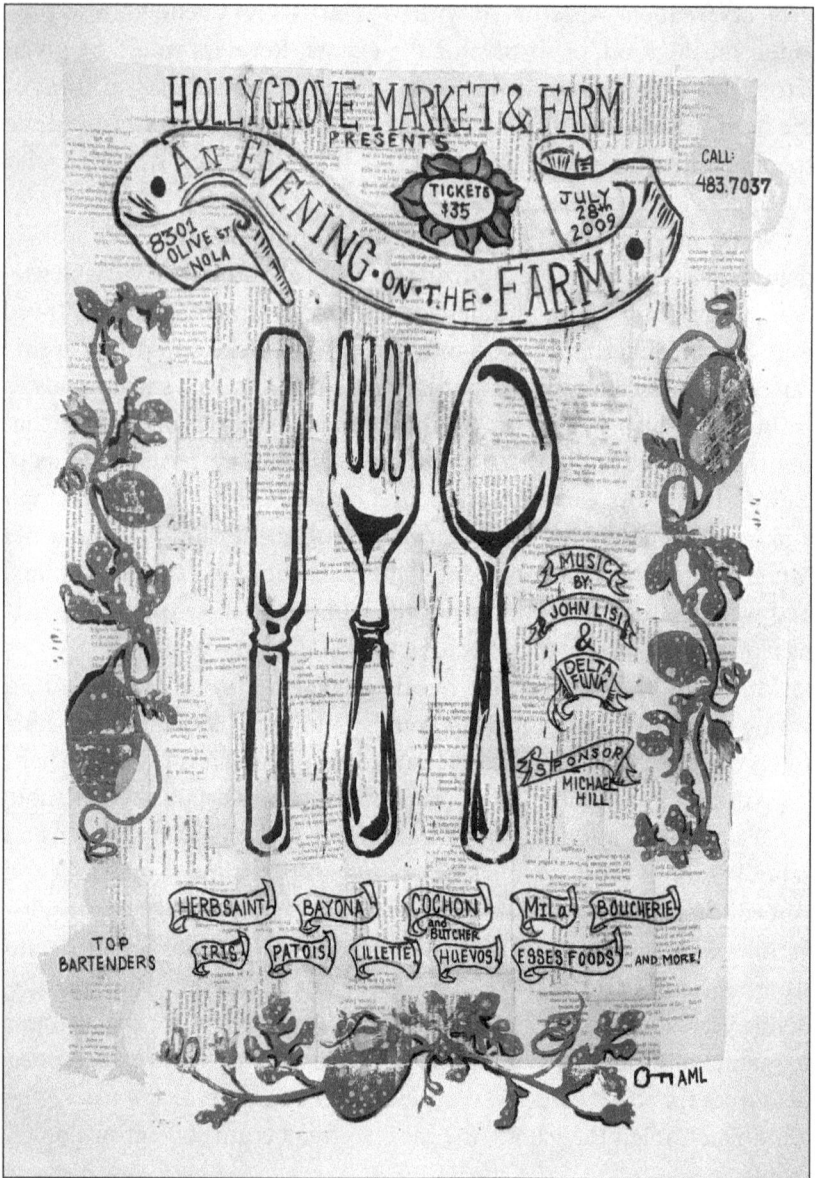

Figure 4.3. Poster for the "Evening on the Farm" event at Hollygrove Market and Farm on July 28th, 2009. Photo by author.

the new generation through their exposure to gardening. But the programs' implied connections between individual choices and life chances largely overlook the structural injustices that underlie persistent racial and class disparities in health and education outcomes. The extent to which these gardens actually make a difference in young people's lives—beyond providing them with a much-welcomed chance to spend time outside and gain some familiarity with the process of growing plants or caring for small farm animals—remains an empirical question that has not yet been examined critically.[12]

Programs aimed at older youth or young adults took a slightly deeper approach to engaging them in urban cultivation. From early on, the founder of Grow Dat Youth Farm viewed the organization within the framework of social justice, with a focus on youth leadership and addressing social disparities and trauma experienced by the city's low-income Black youth before, during, and after Hurricane Katrina. These youth-oriented programs used cultivation as an effective tool to tangibly situate food justice conversations while simultaneously providing concrete and physical tasks that built students' sense of accomplishment. I observed the young adults enrolled in Grow Dat Youth Farm and Our School at Blair Grocery articulate their knowledge of food justice at various public events. It was clear that the young people were not merely reciting memorized statements, but rather drawing their own connections between food insecurity, health disparities, systemic racism, and capitalism. The programs clearly left a lasting impression on how these young people saw the world around them and their own relationship to it. Even so, it is difficult to measure the long-term "impacts" of such programs; it takes time for these new skills and knowledge to manifest in young people's life decisions, and there are too many additional factors in their lives that shape their values and trajectories. These programs nevertheless carried on, with organizers maintaining hope that they would ultimately have some positive impact on young people, even if that impact isn't in the form of growing plants.

Even when they did not organize specific programming for kids, many growers found themselves interacting with and accommodating young people in their day-to-day practice. Neighborhood children of all ages would come over to cultivation sites to hang out after school or on weekends, asking to play with the chickens or to help out in the field.

They gathered at the garden, often because there were no playgrounds or afterschool programming nearby, or because other adults could not watch over them. Most growers rolled with it, after checking with kids' parents and guardians and getting their approval whenever they could. For Zach, taking on this role was a part of what it meant for CRISP Farm to be part of its Upper Ninth Ward neighborhood. When asked about his relationship with his neighbors, he responded:

> I mean, Miss Keesha's kids come by whenever they can, she loves it, they're out of her hair. The kids over across the street come by. Quinn, he lives down the street, his little brother Jay's down there. I mean, there must be fifty kids that I know their names, and there's something cool about taking somebody like Autumn, who I think is twelve, and saying, "Alright Autumn, you're in charge of Chantel," and then asking, "Ok, the rule is you have to be over eleven or accompanied by someone who's over eleven," and I put them in charge and then I go do what I need to do.

The children who came to the cultivation sites to hang out self-selected themselves. They were often particularly interested in working and playing with the chickens. As Zach's comment indicates, welcoming these children into the space facilitated the garden's rapport with adults in the community who might not otherwise have been interested in or known about the cultivation projects. In a few cases, the growers tried to involve the children in actual cultivation activities by assigning "jobs." But unlike the formal programming of Edible School Yard or Grow Dat Youth Farm, these remained informal and open-ended positions, partly to accommodate the varied levels of engagement that these young people sought out.

By meeting people where they were and letting these young people come and go as they wish, these growers inadvertently created a type of social space that sociologist Ray Oldenburg calls a "third place" for the young people in the community.[13] Oldenburg described third places as places, such as cafes and bars, that are neither home nor work, where adults can be free of the obligations associations with those two social spaces. The original theory focused on urban adults seeking a "sense of community" via the casual interactions of drinking and conversation. What distinguishes cultivation spaces as a type of third place was not who is being served (younger people), but how these spaces created organic interactions between these children and the growers as they each discovered their mutual interests and benefits. Though the number of children who regularly came to CRISP Farm was small, their presence suggests the farm's prefigurative role in providing a type of third place that had not previously existed for these young people.

Still, it is important to recognize that these urban cultivation sites were operating in neighborhoods that experienced structural inequality, and many residents were processing both individual and collective trauma.[14] Many of the young people that the growers worked with, formally or informally, exhibited signs of emotional and physical distress that resulted directly from their exposure to the storms and the flooding, as well as from lack of consistent and sufficient treatment for their challenges.[15] Multiple growers shared anecdotes of having to respond to individuals, young and old, exhibiting post-traumatic stress disorder (PTSD) symptoms. One grower shared a story of a young man who used to come to her cultivation site after school to help out on the farm. The grower knew of this young man from a few years prior, when she learned that he had stolen her dog and beaten it to death for an unknown reason. The same young man one day came with a rooster that clearly had been used in cockfighting and presented the bird as a "gift" to the farm. The grower decided to release the bird on the other side of the neighborhood. When he learned of this, the young man became enraged; he claimed that he had been making hundreds of dollars with the bird and threatened to kill the growers on site, though he did not act upon the threat in the end. While most of the young people who came to the farm did not engage in such extreme behaviors of distress and disturbance, working in a community suffering from collective trauma posed a distinct challenge to the growers, most of whom were not trained to work with individuals with such conditions. Other growers recounted working with young people who showed dedication and commitment at the garden but then became subject to the criminal justice system or had moved with their family out of the neighborhood without any advance notice.

In a city where being an urban farmer or engaging in urban cultivation was not a common sight, simply doing the work of cultivation in a publicly visible space could be prefigurative, especially in communities where residents had not readily embraced the new iteration of "urban agriculture" or the "alternative food movement." Jerry recognized that youth engagement with ACRY's programming was not consistent for various reasons, but he nevertheless hoped that it could have some pre-figurative impacts on these young people:

> It's changed a little bit. Some of them still come around. And even though some of them who don't come around, those, when I see them, they're still excited to see me. And, you know, at the end of the day, I was a positive experience in their life. And now they know that a human like me exists. Like whether they ever remember my name, it don't matter. They will remember the experiences that we had. And I think that will have a long-lasting impression on them.

The idea, according to Jerry, is that the simple act of exposing young people to food-growing and community engagement could present them with alternative visions of what they could do in their own lives, even if they were not interested in pursuing gardening, farming, or working with food explicitly. What it would mean to them in the moment, or in the future, however, was not something he could dictate or see to fruition.

Cultivation work kept growers working on site daily, and the outdoor, public nature of their work made them very effective in acting as what urban theorist Jane Jacobs described as "eyes on the street." The people who fulfill this role in urban public space maintain a sense of social order in neighborhoods where most individuals value privacy or otherwise keep to themselves.[16] Unlike in New York City, where Jacobs studied how urbanites formed a particular sense of community in the late 1950s, long-term New Orleanians typically know their neighbors because of their generational residence in the community and the local custom of greeting each other from the front porch or on sidewalks. But in the years following Katrina, in areas where the pace of rebuilding was slow, both the growers and the residents were at risk of experiencing opportunistic theft and vandalism, with occupied units scattered across blocks with limited to no protection from law enforcement. The growers in these areas stepped into this void by keeping an eye on the street because of their regular, outdoor presence.

Jamal described his relationship with nearby residents as generally positive, even though the neighbors were not necessarily involved in the day-to-day operations at his cultivation site in the Lower Ninth Ward. He sensed that most of them preferred his presence to that of the developers who bought empty or blighted properties, just to hold on to them in case the property value appreciated. He described his relationship with nearby residents as being reciprocal without getting deeply involved:

> We used to have people come scrap, steal shit, take plants. I used to go nuts, go scouting in the neighborhood like, "Who got my purple basil?" Because no one else is growing purple basil. And then I realized, "Why am I out here? Wait a minute. If you want the basil, take the basil." . . . People [in the neighborhood] know how to grow, they grow better than me. So people also saw the work we did. I also chased down some construction workers who tried to jack an AC unit (from a nearby house). Because home invasions [and theft] have happened a lot in our neighborhood, like [for] copper [pipes and wires]. So this is just called neighborhood policing. You [people in the neighborhood] are at work during the day. You're here during the night, I'm not. [If] you see a bunch of shit, I'm not asking you to go in there, but like if you see a bunch of bullshit, you don't have to say something, just let me know.

Jamal's site didn't have any fencing, which made it a target for theft. The homes of nearby residents were also targeted, and this created a tacit understanding of the mutual benefits of looking out for each other, as Jamal described. But he also justified the plants' removal, or some neighbors asking to share seeds and soil, as a part of what he provides to the community, even though he was operating his cultivation site for commercial sales.

The interdependence between growers and their neighbors became the basis for mutually beneficial, if not deeply engaged and trusting, relationships. As Jamal pointed out, the aim was not necessarily to perform vigilante acts of community policing, but to provide some preventative and informational resources to each other. When working in this way, the growers could act as what Jacobs called the "public characters" who connect strangers living in a diverse urban community.[17] Public characters foster indirect social capital among these strangers that could be activated as needed, even as the strangers value their privacy. The growers' familiarity with various neighbors allowed them to connect or share casual information about who was back in town or what was happening at a particular property, gathering and sharing such gossip through day-to-day greetings. For instance, when discussing his relationship with the neighbors, Zach described that he had made connections to the people in the community through his garden:

> Bernard and Veronica, who live in that corner house, whenever we have an event, they fry fish and they're able to make a lot of money. We had an event three weeks ago with a couple bands and vendors out here and they fried fish. So, what happened, too, [was] Don over there ran the electricity for the stage, people didn't know he was an electrician. And the neighbor right over there fixes tools. It's something, yeah you're exactly right, brings all the neighbors together. Doug, who lived with Patrick for a while—that house right over there, was painted by him. Nobody knew he was a painter until they were all together. And that's community, right?

The official purpose of the event was to bring neighborhood residents to the garden to introduce CRISP Farm, along with others from across the city that supported the project. But in the process of putting the event together, Zach saw that the garden ended up connecting the neighborhood residents in unexpected ways.

The growers' relationship with one another similarly remained a loosely connected network of what sociologist Mark Granovetter calls "weak ties"[18] rather than forming into a collective identity; this reflected the diversity in their locations, aspirations, and forms of cultivation, in addition to their varied backgrounds and identities. They worked mostly independently from one another, even as their numbers rapidly increased across the city during the redevelopment period. Their relationships with each other ranged from close partnership and collaboration to polite recognition without direct interaction. While there were a few attempts to create a space for organizing the growers and others interested in urban cultivation and local food issues broadly, these conversations did not materialize into actions.

Foodshed Roundtables was one of the first organized attempts to foster food justice conversations at growers' homes or in their cultivation spaces. I attended one of these dinners in 2011 and observed over thirty individuals gathered in the evening at a vacant lot in the Bayou St. John neighborhood. The participants were overwhelmingly white, the majority in their twenties. After guests helped themselves to empanadas provided by the event organizers, we were asked to select one of the four tables based on the topic we wanted to discuss: What is a movement?; What is our vision?; What are our values?; and What's up in our food system? I chose "What's up?," and Cory, as one of the organizers, set the tone by sharing that the aim of the event was to engage in an experience-based conversation about the food justice movement in New Orleans. Six other participants at my table, including two growers whom I knew, spoke animatedly about the negative environmental impacts of large-scale monoculture and the distances food traveled from the point of production to the point of consumption. As the tables were sharing summaries of their discussions, I sensed that Cory and the other organizers were trying to emphasize points that were more aligned with food justice concerns, such as respecting the history and culture of the work or addressing structural racial and class injustices. The participants, however, appeared to be more comfortable discussing alternative food network talking points than food justice concerns, for which we collectively raised more questions than answers during the conversation. The Foodshed Roundtables stopped happening by the end of 2011, about a year after their introduction.

Efforts to organize emerged again in 2014, when the city announced revisions to the Comprehensive Zoning Ordinance. A group of growers mobilized in response to the announced regulation for the new "urban agriculture" land use category, which included not only the extensive requirements for soil testing discussed above, but also provisions for food handling and beekeeping that did not reflect the realities of growing in New Orleans. The new policy employed language implemented in other states, and the growers had not been consulted until the initial language was already drafted. This lack of consultation occurred despite growers' attempts to recommend guidelines to proactively shape the new ordinance. About a dozen growers continued to put pressure on the city to amend the regulation. These efforts became the locus of an organizing movement for these growers, who formed the Greater New Orleans Growers Alliance (GNOGA) in 2012, a collective of farmers in and beyond New Orleans to organize around policy advocacy at local and national levels.

Conclusion

Prefigurative urbanism is an "ends-effacing" form of prefiguration that prioritizes changes that can be implemented now, rather than working toward a fixed end in the distant future.[19] The ends-effacing nature of the practice allows actors to innovate and adjust their practices as they discover what works and what does not, or as they encounter new possibilities they did not originally anticipate or consider. Regardless of their initial aspirations, actors who engage in prefigurative urbanism approach their practice with individualized pragmatism. Abstract theories or visions only matter to the extent that they can be put into concrete actions. In fact, in prefigurative urbanism, the actions clarify the reasons, not the other way around.

Prefigurative urbanism is exploratory, experimental, and adaptive. As actors of prefigurative urbanism, the growers disregarded and blurred such existing social categories as for-profit and nonprofit, or public and private, as they tried to make urban cultivation *work* within the existing system of urban political economy and social norms. Unlike prefigurative politics, whose aim is to disrupt, undermine, or surpass hegemonic institutions and norms in enacting social change through direct action, prefigurative urbanism is not mobilized around political grievances or a collective vision toward these changes. Instead, it is activated by the actors' need to enact efficacy in publicly recognizable forms that yield observable outcomes in a relatively short period of time. If the existing system is disrupted, the disruption is a byproduct of the practice. The actors respond to social or political barriers by either disregarding them or finding solutions outside of the existing system rather than overtly resisting or advocating for change. After all, the point is to keep on keeping on.

Can prefigurative urbanism be transformative, and if so, to what extent? The changes enacted through prefigurative urbanism tend to be hyperlocalized and specific, such as operating a five-customer CSA in one neighborhood or providing a space for a dozen children from the neighborhood to hang out after school. These forms of prefiguration are not insignificant, but they do not catalyze broader societal change, at least not directly. As some growers insisted, their actions were "demonstrative" of the alternative potentials for land use, food systems, property ownership, and career choices. Even though they are experienced at a very small scale, the public visibility and tangible indicators of these changes constitute a convincing argument for what *could be*. Prefigurative urbanism actors are cognizant of the constraints posed by the limitations of existing social systems, but find hope in their self-efficacy to enact forms of alternative urbanism at the scale that they knew their current capacity afforded.

Even when growers made a strong commitment to persist, sustaining urban cultivation as a primary occupation, whether as a for-profit or a nonprofit entity, proved challenging for most of the growers in this study. It is worth noting that not all growers saw urban cultivation as their lifelong career, except for the growers with *urban cultivation expansion* aspiration, as well as for retirees. For the younger growers, sidestepping the question of whether this was their new permanent career left all their options open, including the possibility of departing the practice in the future. With its commitment to the now, and with constant adaptation and reconfiguration, prefigurative urbanism can be difficult to sustain over the long term, as the next chapter demonstrates.

5

When the Garden Goes Fallow

Trajectories of Prefigurative Urbanism

In 2018, I walked up to the gate of what was once a thriving urban farm in the Lower Ninth Ward. It still had the yellow metal sunflower on it, but that's about the only thing that was immediately recognizable from when it was the headquarter of Our School at Blair Grocery. The building of the former neighborhood grocery store was still standing, and if you stepped closer to the gate you could almost make out some of the murals and words of empowerment painted in bright colors on the walls, though they had either faded or been covered by the overgrown plants. Stacks of old tires and barrels lay scattered on an empty lot across the street, where I once saw a group of volunteers erect hoop houses during a weekend workshop. No more rows of vegetables, chickens picking through piles of compost, goats bleating, hydroponic structure whirling, or young people working in the heat or hanging out in the shade. It was quiet. Not one car or pedestrian passed by while I walked along the site, trying to find the right angle to capture the scene in a photo.

Our School at Blair Grocery was just one of many former urban farms and gardens that I found blighted during my visit to New Orleans in 2018. I learned that at least a dozen lots terminated cultivation activities since 2015. Some growers lost access to the land and others walked away from their cultivation practice or from the city altogether. Five years had passed since I had begun this project, and I was left with more questions than when I started. What makes urban cultivation sustain, and when and why do some cultivation projects discontinue?

Cultivation projects ended for various reasons, and including both external and internal factors (Table 5.1). The termination of some of these cultivation projects leads us to contemplate the variant long-term trajectories of prefigurative urbanism, and what it means for it to be sustainable or even successful. Because of its ends-effacing nature, defining

Figure 5.1. Former site of Our School at Blair Grocery in 2018. Photo by author.

an "ending" of a prefigurative urbanism project is not nearly as straight-forward as identifying its beginning. The termination of a project at spe-cific location does not necessarily mean the end of the growers' work, nor does the departure of a grower from a project or a space mean the end of that space being cultivated if another grower steps in to take over. Projects that started as a form of prefigurative urbanism could evolve into a social movement or prefigurative politics, changing the orienta-tion and missions of the projects even while the day-to-day practice re-mained mostly the same.

TABLE 5.1. Factors Impacting Termination of Urban Cultivation as Prefigurative Urbanism

External factors	Internal factors
• Loss of cultivation space	• Financial insecurity
• Community changes	• Mission creep
• Market instability	• Limited skill and capacity
• Formalization and Regulation	• Life stages
	• Burnout

Factors Impacting Termination of Urban Cultivation Sites

Prefigurative urbanism practices are inherently precarious, due to their defining characteristics: they are ends-effacing, direct action-oriented, experimental, and not centrally organized. The resources needed to sustain the practice may also differ from those needed to launch, especially when the project's aspiration and scope transform or expand over time, but identifying and accessing these resources requires additional energy and attention from the grower whose priority is to *do something, now*. Unlike conventional social movements, prefigurative urbanism does not organize around collective grievances or identity, and this fact increases the risk of termination when the initial actor no longer finds value in the act of prefiguration or feels that whatever action they had taken on the project was sufficient.

Social movement scholarship uses the concept of *abeyance* to explain how a movement that may appear less active during certain periods of time, especially under hostile political and social environments, continues to mobilize to sustain its work.[1] This theory argues that movements in abeyance do not *end* but instead adopt different internal and external strategies to persist. Activists continue to push for social change as the political and social contexts around them transform, sometimes as a result of their own efforts. Individual activists experience shifting levels of engagement with the movement throughout their life course, in connection with the movements' own ebbs and flows, but also reflecting their own life stages.[2] This is because one of the core motivations for activists' long-term commitment to a social movement is the formation of a collective identity associated with that movement.[3]

Prefigurative urbanism similarly persists through experimentation, exploration, and adaptation, as the previous chapter illustrated, and their ongoing involvement reflects their intent to continue a practice that they see as enacting immediate changes. But these actors are typically not motivated by their desire to participate in collective activity or identity; rather, they opt to take on actions individually to enact their own visions of change. As a result, the trajectory of prefigurative urbanism rests largely on individual actors' orientation toward the practice. It persists as long as they find it worthy of their time; in what form it continues is less important than the fact that it persists. When the practice terminates, participants experience that cessation as an individual decision rather than a collective loss.

This individualized approach is both the unique strength of and the weakness in prefigurative urbanism. Individualized practice means that the termination of a project does not affect other practitioners. There is no collective setback. At the same time, the lack of a collective identity means that the actors must find and retain their own motivations for continuing their practice. Social movements draw their strength from the "collective hope" that sustains motivation among individual activists, and a movement that is overly utopian or pessimistic struggles to retain its members.[4] By contrast, in prefigurative urbanism, actions sustain individuals' "willful hope," which is rooted in pragmatism. In other words, the termination of a project of prefigurative urbanism does not simply reflect the lack of interest or hope among the actors; it is the result of other conditions that pose challenges to their ability to retain hope through direct action.

Prefigurative urbanism enacts willful hope to activate the sense of individual efficacy with wider social implications. Willful hope is essential for prefigurative urbanism to persist, and McGeer warns that actors driven by willful hope risk despair when they lose confidence in their capacity to enact change.[5] Whenever growers faced external or internal challenges, what kept them going was their inclination to remain hopeful. Growers did not necessarily see their efforts as a failure even if a project ended; after all, they had enacted an alternative way of growing food and using urban space, albeit for a limited amount of time. Challenges were to be expected precisely because they were trying to manifest something that did not yet exist. The exception—the cases where growers perceived their projects to have failed—occurred when growers exited urban cultivation or the city altogether because of burnout. Burnout, as the growers described the experience, was not just a feeling of physical or mental exhaustion, but a culmination of internal and external factors that made them feel that their efforts were no longer producing an outcome they found meaningful. In other words, prefigurative urbanism actors truly ceased their engagement with the practice only when they felt that it no longer sustained their sense of hope.

Loss of Cultivation Space

Land access and tenure are major concerns for the long-term sustainability of urban cultivation. When growers who had access to multiple lots lost a space, they could reorganize their projects and scale them down. But even then, losing a space could significantly disrupt their operations. The situation was more tenuous for growers working in a single plot, and growers' land-access insecurity intensified when the area was under speculation for development.

In 2018, Shannon and her growing partner were displaced from the space in Mid-City that they had taken over from another grower. The space had been tended by a long-term gardener since 2008, but when that gardener's health conditions declined she passed the space on to another experienced grower before Shannon took it over sometime around 2015. But almost immediately after she began cultivating the space, Shannon got caught up in litigation over lot access due to legal uncertainty over who actually held the right to lease the land to the growers. With the help of a local attorney specializing in urban agriculture issues, Shannon was able to remove herself from the litigation, but she suspected that the contention over ownership reflected anticipation that property values were rising in the neighborhood: "Once development happens, it's starting and people are doing landgrabs. Obviously the farmer, the grower, is almost always going to be the person who loses out, because almost never do they have the resources to begin a long legal battle like this." This was not Shannon's first or the last time losing her cultivation space in New Orleans. As described in Chapter Three, another lot in Mid-City that she and her growing partner were leasing was reclaimed by the owner, who proceeded to use the land themselves. There was also a community garden that she and fellow gardeners were once told to vacate when the owner attempted to sell the property, only to be given to another grower when the owner realized he could not sell it due to the amount of back taxes he owed on the property.

Starting around 2015, developers began acquiring properties that used to be gardens and farms, then would raze the growing facilities only to do nothing with the land. Nearby residents and the growers who used to tend the land had to endure the sight of the space sitting vacant for months or years and often become blighted again. Cory recalled how common this process was:

> It happened to all of the lots we got kicked off. We got kicked off sequentially of one lot, you know, that we were technically squatting on, but someone bought it [and then] they kicked us off. But [they] didn't do anything with it, and also didn't maintain it. Another lot was sold off that we were on consensually, well, there was an agreement. Same thing [happened], you know? And now this final, biggest lot [the access to which was lost in 2020].

Evicting urban growers in the name of real estate speculation is not unique to New Orleans.[6] The uneven pace of redevelopment across the city meant the risks of losing access to cultivation projects varied dramatically based on where they were located, though the speculative nature of developers' land-grabbing behaviors meant growers typically lost these spaces long before actual development began to change the economic and social makeup of these neighborhoods, if it ever happened.

By 2015, there were many signs that organizations and agencies that had been leasing their lots to growers were becoming more selective about which lots they would continue to lease or even make available for purchase. They were granting access to their properties in the name of supporting sustainable land use, but they undoubtedly also benefitted financially by having the growers taking on the cost of managing otherwise blighted properties. One of my interviews in 2014 took place shortly after a project had received a notice from NORA that it would not be offered the opportunity to purchase the land, which was the option that had been explicitly stated in their initial lease agreements. The garden manager read me the email they received from NORA:

> At this time we are prepared to renew the lease agreement on the current properties. We won't sell them at this time, nor will we lease additional properties. It is our desire to see these properties properly maintained and grown into fruitful urban gardens or farms. Please continue this effort you have started and let me know if I should draft the appropriate documents.

The terms of a grower's lease, even when leased from a state agency or a large nonprofit organization, remained somewhat subjective when it came to determining whether the lease would be extended beyond the original agreements. The judgement was often based solely on the owner's assessment of the quality of use or the property management conditions, putting growers in precarious situations when property values began to rise rapidly in some areas of the city. In part because leasing organizations had a relatively high turnover in personnel, they lacked the organizational memory to honor the handshake or verbal agreements under which the growers initially entered leases.

Even Habitat for Humanity, whose HUG program initially had a reputation of being a straightforward, hands-off lease program in contrast to that of NORA, began responding to developmental pressure. When we initially interviewed Margee in 2014, she had been leasing a space through the HUG program since 2012, making her project one of the program's first successful urban cultivation leases. By 2018, when I conducted the follow-up interview, she was facing uncertainties about that lot. She had been up for the lease renewal in 2017 for another five-year term with HUG and asked the organization whether it was worth investing in some infrastructure for the long term on the lot:

> I said [to Habitat for Humanity's HUG program staff], "I'm looking to invest in irrigation, just wanted to make sure that we are still on track for me to get that second five-year lease," and they said we're good to go, and so I invested in a line up for water and a water meter, which is a pretty substantial financial investment. . . . So they verbally guaranteed that I would have another five-year lease, and then I kept trying to come in to sign that lease, and was kinda getting pushed off or not answered. And eventually . . . they called me in February [of 2018] and they were like, "Hey, just heads up, we need you off the lot by March."

She tried to negotiate and make a case, but with only a verbal agreement, her options were limited. At one point Habitat for Humanity proposed that Margee buy another "comparable lot" in the same neighborhood so that they could "swap" the lot ownership. But with the property values in the St. Roch neighborhood rising rapidly, Margee had a difficult time finding a suitable lot; when she finally found one, the owner remained cagey and unresponsive about finalizing the sale. Habitat for Humanity never fully explained to her why the lot she was leasing was now slotted for housing development, when they had plenty of other lots across the city where they could have built a home, but Margee suspected that it had to do with the neighborhood's gentrification. Though she was hoping that the process would drag on, allowing her to continue growing where she had been for over six years, eventually a home was built and the property was sold in 2020.

Unique to the gray areas that these projects occupied, the growers' prefiguration of the alternative potential of these spaces could not compete with conventional, capitalistic property values. The individualized nature of growers' practices distinguished their cultivation projects from community gardens, whose gardeners have been able to collectively define and advocate for the gardens' public benefits, in order to secure long-term land tenure.[7]

Community Changes

Gentrification and changing neighborhood demographics made particular impacts on cultivation projects with *community rebuilding* aspirations. Joel first operated a gardening project in Central City called Delachaise Gardens, with involvement from people in the neighborhood. The project occupied two lots that had been abandoned and razed after Katrina, now with an absentee owner. Joel and his gardening partner "took advantage of that and took it over as a garden," turning it into "manicured, good-looking gardens" with chickens, signs, and two little hoop houses. For a time, their garden provided affordable produce to the long-term residents and created a space for the neighborhood youth to come hang out or work. But Joel could read the "writing on the wall" when the houses near the garden began to be sold to developers who would then flip them:

> The neighborhood had been changing in terms of people getting some of these houses that were abandoned, and there were homeless people who were living in them, who were friends of ours and friends with the garden. And then once [a house] got bought off and flipped, they moved. And then some people got arrested, some people died. The neighborhood

had changed. Pretty much, it coincided with the need for the garden not really being there anymore for that. . . . And then at certain point there's only *x* amount of people still left in the neighborhood who are originally part of the garden. The writing was also on the wall, so we just kind of let it go.

Eventually the owner of the lot, who had initially ignored the presence of the garden, decided to raze the lot in preparation for selling. Since Joel and his growing partner never formally established access to the space, and seeing that the community that they had been working with was being displaced, they decided to move on. This was when they relocated to Central City to start Paradigm Gardens as an entrepreneurial cultivation project.

Neighborhoods could also change in the opposite direction. As Joel and his growing partner were transitioning from their first garden to Paradigm Gardens, Pamela was in her third year of a multiyear soil-remediation process for Sun Harvest Kitchen Garden in the same neighborhood in 2012, when she noticed an increase in the number of unhoused people living in the immediate vicinity of the garden. She had misgivings not only about their number, but about how they were starting to use the space without permission. She recalled:

My suspicion was that they were coming from other cities and finding New Orleans's climate favorable. Because prior to that third year, the local, established population never came into the garden, you know. They would just pass by and admire. The garden is one block away from the Mission. And they'd just stop by for conversation or ask if there was anything they could do to help or say how beautiful it was. But I really noticed in that third year that there was something else going on.

She began noticing that people's belongings were being left in the garden, in wooden bins that she had used for storage and composting. She continued:

So, I started having to negotiate, sort of this dance of respect for not touching their things. And eventually within a day or so they would remove them. But then, it started getting more and more unsafe. I started seeing signs of people defecating in the rear of the garden. Not directly in the garden but along the borderline. And then over time I would approach the garden very early in the mornings and find someone still sleeping there, you know, that sort of thing.

Because she intended for the produce to be used at the restaurant, she felt that the space could not continue on its current trajectory. She wrote to the city, with support from Café Reconcile, in hopes of getting some assistance in managing the increase in the unhoused population in the area, likely due to its proximity to the Mission, a nonprofit organization that provided services to unhoused population. The city responded with mostly symbolic gestures, including a visit to the site and a temporary removal of the unhoused population from under the expressway a few blocks away. In the end, Pamela decided she could not go forward with the project because of the "liability concern" about the quality of food that was to be served at the restaurant down the street.

When I caught up with Jenga in 2020, she shared that her organization was about to lose access to a house adjacent to one of the gardens that they had been using for over a decade. There was an abandoned house next to the garden that Jenga's Backyard Gardeners Network tended in the Holy Cross neighborhood of the Lower Ninth Ward. In 2009, Jenga and the gardeners had learned that Preservation Resource Center (PRC), a local nonprofit organization focused on historical architectural preservation, owned the property. When the gardeners approached the organization, PRC agreed to renovate the exterior of the building for the Backyard Gardeners Network, if they took on the internal renovation. Upon completion of the renovation, PRC offered to transfer the ownership of the property to the Backyard Gardeners Network, but Jenga felt that the organization was still too small and young to own a property, so she reached out to the Holy Cross Neighborhood Association (HCNA) and it became the legal owner of the lot. Jenga and the gardeners continued using the house for storage, training, and other community programming purposes, but over the decade the neighborhood's makeup shifted with an influx of white transplants while the long-term residents' rate of return remained slow. At the beginning of 2020, Backyard Gardeners Network approached the neighborhood association to inquire about acquiring ownership of the house, when Jenga felt that the organization was ready for the next step:

> We had been using it for ten years, and we just saw a lot of value in having ownership of it. We could leverage that for various things we could use it for. We talked about, in the future, having gardener exchanges, you know, we could actually have someone stay there for a certain period of time while they're in New Orleans. Or if we needed to earn revenue and lease it out for a certain period of time as a shared space with us. There's just different options. And so we approached them with that proposal and they actually came back and said, "No, we will not transfer ownership to you. Also, we want this property, and so you need to leave." (That) is what they said. [Pause and dry laughter.] "You need to leave and get your stuff out, and we want to look at how to use this property for the benefit of the community."

After several conversations with the neighborhood association, it became clear to Jenga that HCNA was mostly unsympathetic toward her project. So, Jenga told the gardeners that they were no longer able to use the space and that they needed to move the stuff out of the house. In retrospect, Jenga recognized that her organization could have been more proactive in making the neighborhood association aware of the history and Backyard Gardeners Network's connection to the space and the community, even through signage at the house or by attending the association meetings. But for various reasons, she gradually stopped engaging with HCNA over the decade, just as the neighborhood was experiencing gentrification. Backyard Gardeners Network remained focused on community engagement, especially with the long-term, Black residents; meanwhile, this priority resulted in the project failing to retain a strong relationship with the neighborhood association, whose representation was changing in response to the community's gentrification:

> One thing that we did do in the past in terms of communication that was above and beyond what organizations usually do is send our youth interns out into the community. And they would deliver flyers door to door, you know, within a certain radius of the garden to let people know.

So it's like we kind of overdid that and, then underdid [communicating with the neighborhood association]. It's lessons learned, you know. You're not always perfect, but I feel like at the same time, it's really not fair for these other people who all have been in this neighborhood [briefly] to just come in and make assumptions and decide what's not working without even investigating or without trying to understand or trying to genuinely offer help, if they really do want to help. Yeah, that's hard for me.

In hindsight, too much focus on the needs of the long-term residents, or the immediate capacity of the organization, shifted her focus away from planning to maintain long-term access security to the house as an organizational asset. Even then, she did not expect the neighborhood association to respond in such dismissive ways, precisely because of the work she had been doing in the community over the decade following the 2005 storms. This did not entail loss of the cultivation space itself, since the garden next-door was on a more secure, long-term lease through Parkway Partners, but retaining the access to the house would have been a potential asset in catalyzing a resurgence of the organization through new financial and programming capacity.

Jenga recognized that the organization's capacity had been on the decline by 2020, and that in the end it might have been for the better that they did not retain the access to the house. Nevertheless, the way HCNA disregarded their presence and work in the community during the process left her feeling bitter about the disrespect toward Backyard Gardeners Network, as it made clear the association's selective support for newcomers and their interests. As her comments indicate, gentrification of the neighborhood was not only impacting the property values but also the community's relationship with the cultivation projects. Jenga noted that her "biggest concern" was the "history being erased" and rewritten from the perspective of the newcomers about how they saved the underutilized house and made it better for the community.

The contrasts among Joel, Pamela, and Jenga's experiences show how the effects of neighborhood change can be highly localized and can extend beyond property values. Changes in who lives and hangs around the cultivation sites changed the growers' day-to-day practice, because prefigurative urbanism is enacted in publicly visible and accessible places. The spatially fixed nature of urban cultivation restricted the growers' ability to move their projects in response to community changes, especially when they were motivated by the *community rebuilding* aspiration.

Market Instability

In August 2015, Good Eggs abruptly announced the closure of their operations in New Orleans, New York, and Los Angeles. The organization would retain only its original San Francisco location, citing the difficulty in making the "last mile" of the food distribution system work elsewhere. The decision highlighted the unique nature of local food systems that may not scale up easily or be transferrable to all urban areas. It also was a blunt reminder to New Orleans's entrepreneurial growers of the risks of relying too much on external opportunities to sell their products.

Caroline was selling "$700, $800 worth of produce a month," within a few months of starting Grow Me Something, and Good Eggs became an important part of her sales. When it closed, it impacted her retail sales immediately and significantly:

> I had two customers that were buying pretty much everything I was growing, but like I said, I couldn't produce enough by myself to support myself. Scaling was really a problem. I was growing, like, lettuce greens, things that at the very best, they're worth $8 a pound, and a pound of lettuce is a lot. Good Eggs went under in the summer of 2015, and I sort of knew the jig was up. But it took me until the following January to really [say], "Okay, I'm not going to do that anymore. I can't make financial sense out of it."

Good Eggs' short stint in New Orleans exemplified the ways in which out-of-town ventures arrived in the city with a significant financial backing from outside, then quickly retreated when they could not replicate prior successes. The company's approach to growth left disrupted markets and social relationships in its wake. As described in Chapter Four, the growers had mostly been cultivating their own individual sales or donation networks during the recovery and transitional periods. While those networks took time and energy to develop, the relationships they built with community members or restaurants created real, or at least a perceived sense of, stability in their long-term prospects for continuity. The arrival of Good Eggs enabled the growers who started their projects a bit later, around 2014, to establish their market reach quickly, but without the local relationships or distribution system that might have ensured their long-term survival. One grower observed,

> And I think the biggest problem here is . . . it's just timing and luck for a lot of farmers. You know, we've never really had a permanent nexus here to connect growers to each other or to the market from the people who would buy our products. [The markets] come and go and they're here for like half a year to a year. . . . So whenever [market arrival and closure] happens, you see sort of this potential to actually pull off urban farming in the city disappear, and then back again. Most recently it happened with Hollygrove, which is the one semi-permanent, somewhat dependable bastion for that.

Hollygrove Market and Farm's closure in 2018, after ten years in business, resulted in large part from financial mismanagement, including the operators' failure to negotiate for a more secure and reasonable lease agreement with the landowner.[8] A group of growers then attempted to negotiate with the landowner to carry on the operation as a growers' cooperative, but they, too, failed. The lot was eventually razed and sold, and remained vacant as did so many other former cultivation sites.

The rise and the fall of a market that initially aimed to increase local access to food in a neighborhood considered to be a "food desert" was its own form of prefigurative urbanism. Its ultimate demise reflects the challenges of scaling up and sustaining a practice that sought to enact an alternative future but that only made sense in a particular time and place. Yet Hollygrove Market and Farm also played a crucial role in incubating, training, and creating a network of growers during its decade of operation. Thus, while the termination of both markets produced direct, negative impacts on local growers' capacity to market their products, the prefigurative nature of Hollygrove Market and Farm afforded the urban cultivation scene with a place for collaboration and exploration prior to its demise. This stands in contrast to the Silicon Valley vision of Good

Figure 5.2. Hollygrove Market and Farm 2010 (left) and 2023 (right). Photos by author.

Eggs, whose implementation neither left room for collaborative adaptation nor contributed to ongoing prefiguration actions.

Formalization and Regulation Enforcement

The city and its regulatory agencies remained mostly uninterested in urban cultivation for much of the decade, for better or for worse. Previous chapters illustrated how bureaucratic inefficiency and inflexibility created challenges for the growers when they tried to establish new cultivation projects during the recovery and transitional periods. For a few growers interested in the policy implications of their practice, this represented a manifestation of larger problems in the existing food system, as well as a lack of understanding of or interest in urban cultivation. Nevertheless, most growers also reported a silver lining in the city's lack of interest in urban cultivation: lack of oversight and regulation.

Toward the end of the decade, the city did take notice of urban cultivation, and "urban agriculture" was included in the Comprehensive Zoning Ordinance revision in 2015 as a land use specification. As noted in Chapter One, the initial soil-testing language was unpractically restrictive before it was revised in response to lobbying by a group of growers who demonstrated that Louisiana State University could not accommodate the kind of testing the regulation required. For many growers, the more meaningful change in 2015 was not about the new language on the books but how the existing rules were being enforced. Rob described the change from his perspective as follows:

> There was land-use stuff happening with farms across the city because, at that point, in 2014, that was about six years into, no, really, it'd been five years or four years into the massive growth of farms throughout the city, whether they were nonprofit, for-profit, community-based or whatever. It had reached a saturation level where the city was like, "All right, we've tried a whole bunch of different ways to get our heads around this, and we haven't been able to yet, so we're just going to start enforcing stuff and stop caring about how it impacts people, how this little garden being on a corner over here somewhere affects the community, we just need to slap them with the fines and be like, 'Hey, get it together or you're done.'"

The regulation enforcement seemed inconsistent and arbitrary, which created speculation about why some projects continued to get away with violating existing codes on fencing or animal husbandry, while others received increasing scrutiny, not just from the neighbors but from inspectors or the organizations leasing the lots. Rob continued:

> It was little stuff. Somebody was driving by and saw us herding the goats, or maybe the goats had gotten out in the middle of the day and they called animal control. Animal control comes out, and of course, they have to file whatever paperwork and all that. But they never did anything about it because they were like, "Everybody's fed. All the goats are fed. They're happy. Everything's good with this. They're in a confined space. As long as everything's working right, we don't have a problem with anything you're doing, and we can't find any laws that are stopping you from doing this, but every time that somebody calls, we have to come out here and do this." We were like, "Well, okay." . . . It was just dealing with little hassles like that just constantly that made us want to focus on getting everything that we were doing licensed and permitted and go through the proper channels to get it done.

Bringing the operation in line with existing regulations could take up a lot of growers' time and energy, especially if their access to land was precarious or an agency was not clear on how to handle a specific form of agricultural land use. The increasing scrutiny also meant they needed to adjust their practices, even when there was little oversight and regulation, especially during recovery and transitional periods. In some cases, growers' efforts to make the project "legitimate" created more trouble than if the growers had continued operating in the legal gray zone, as many of them did for a while. Growers viewed regulatory enforcement as haphazard and arbitrary; it did not appear to display any notable racial biases based on the growers' identities or where they were operating. The city's attention seems to have been activated when a disgruntled neighbor or other growers called attention to a project, rather than by any form of routine inspection from agency officials.

A grower who started an orchard project in the Upper Ninth Ward around 2011 recalled the changes in blight citation practices that started in 2013–14. At the time, she was leasing three lots from Habitat for Humanity:

> Luckily, during the winter, the weeds kind of stay back and you don't have to worry about it too much. But during the summer we were up there every other week trying to mow nine lots and the house that we were taking care of. And it was just, it was impossible to keep up with it. The first two years were actually fine, if they were a little overgrown, no big deal. But [then] that was just, like, becoming such a common thing for people to have their [cultivation] lots up there, that the city really started taking notice. And we ended up getting two citations for two of the lots. And one of them I absolutely deserved. Oh my god, it was so bad. But the other one, like, they were just nitpicking. And that was not my full-time job, so to have to go out and sit and do a citation hearing and to try to get out of these, I mean, it was just, it was crazy.

Shortly before we interviewed this grower in 2014, her organization relinquished these lots back to Habitat for Humanity. By that point, she had planted fifteen fruit trees on the lots and tended them for three years. Even though the orchard itself required relatively little maintenance, keeping up the appearance of the lot to comply with the newly enforced blight ordinance became too time-consuming for a small organization like hers. The grower shared that her organization was considering shifting its approach to planting fruit trees in private yards rather than developing an urban edible forest, even as it continued its original focus on gleaning fruits from existing trees across the city.

∾

These external factors that threatened the sustainability of cultivation projects could have affected any of the growers, but Chapters Three and Four have illustrated how they mostly managed to adapt, whether by innovating new practices or pivoting to other organizational goals. It was when these external factors coincided with internal factors that projects ended or growers left cultivation. These internal factors included financial insecurity, a lack of relevant skillsets, organizational and leadership restructuring, life-stage transitions, and burnout.

Financial Insecurity

It is not easy to make it as an urban grower, especially financially. Unless growers had a grant-funded salary provided by a nonprofit organization or additional sources of income, they found it difficult to make a living. These growers made it clear to me that their interest in cultivation was never about money but instead about doing something to make changes in the lives of others or of their own. Yet when asked about what they expected or anticipated "five to ten years" into the future, the growers expressed cautious optimism regarding the long-term prospects for sustaining their work. Their concerns included access to resources, the dynamic and experimental nature of the practice, and their personal capacity for sustaining the work.

Colleen loved teaching and "the idea of people getting excited about seasonal produce," but she expressed trepidation about her capacity and willingness to stay with urban cultivation:

> Do I want to invest this much time and energy into people who may or may not ever value it with money? . . . I don't like that I can't find a way to make money doing it, and I don't like the idea of chasing grants. I would

rather just be paid for the work that I put into something. . . . An urban farm doesn't make a lot of money. So where do these people who are your students [go to work]? I want to be good enough to where they want to pay tuition. But in reality, at the end of the day, I feel like we all need a little urban farm reforming, you know, it's just like a big dreamy idea, but in reality, what's the product?

Most growers shared Colleen's grievance that agriculture is undervalued by the state and the public, and that urban cultivation is not economically sustainable whether it is for-profit or nonprofit. Eric recalled noticing that customers at the market were willing to pay a lot of money for herbs and flowers but hesitated to spend the same amount for his vegetables. He would say to himself, "I'm in the wrong business." Eric found this to be true for customers regardless of socioeconomic status. It was especially discouraging when the wealthier consumers continued to expect lower prices for the produce.

The growers with skillsets beyond urban cultivation, especially the younger, commercial growers, continued to work in other sectors to supplement their income or to fund their projects. Jordan described the number of jobs he held at the time of the interview as follows:

Yeah I work so many different jobs, my God. I'd hate to say I'm kind of a hustler, you know? Like, there's always one thing or another. Odd jobs, building. I'm actually in a carpentry school right now, thinking of trying to get more carpentry jobs, 'cause the landscaping is killing me. I had to go six weeks, more than six weeks without working because I did dislocate my left shoulder completely unrelated to work. Other thing is teaching, but I don't want to teach in the school system. So I've been tutoring actually. It's the other new hustle to go with that.

The side jobs that growers held were wide-ranging, from teaching, event planning, to consulting on tech services. The growers who did not pursue additional sources of income were mostly retirees, had a family or life partner with stable income, or were getting paid through grants or fellowships such as AmeriCorps Vista programs. It was not unusual for a side hustle to become a grower's main priority over time, affecting their capacity to dedicate the same amount of time and energy to the cultivation project.

This dynamic led Pamela to ask, "One of the things that I see is, there is no pathway to resources. If you're not able to do this out-of-pocket, then what?" She recognized that there were issues with NORA's land access programs, but she also pointed out the growers were "expected to maintain and develop growing operations" with their own means once they finally gained access to the agency's vacant properties. She problematized the uncertainties of and disparities across the long-term viability of cultivation projects, especially when growers had few options for financial stability.

The financial insecurity and uncertainty surrounding urban cultivation reflected how the practice continued to exist outside of normative urban economic and social systems. But in this context, it is notable that growers consistently sought individual solutions to these shared problems. Their priority was to continue growing, rather than to mobilize to demand changes in the resources made available to their practice.

Mission Creep and Organizational Restructuring

The open-ended, innovative, and experimental nature of prefigurative urbanism put many projects at risk of mission creep, whereby ongoing practices became misaligned with the original aspirations for the project. In 2011, the entire staff of Our School at Blair Grocery (OSBG), except for the two main leaders, exited; they accused the leadership of financial mismanagement, poor communication, lack of democratic structure, misalignment of mission and praxis, and exploitation of Black trauma in the community, among other grievances.[9] The letter sent to the two leaders of the organization, signed by the "staff, student body, and the community," demanded action as follows:

We demand: a system of checks and balances in relationship to the structure of power within OSBG, the assembly of a functioning board of directors to include representation from the community we serve and our fundamental values; the ability to engage in collective reflective process allotting us the ability to assess our work; intentional development of OSBG as a school and not just an urban farm; open communication between the many facets of OSBG, dissolving back-door meetings, negativity and slander; a clear understanding of our individual roles and shared responsibility in the greater goal; intentional staff development and training in order to better our ability to perform exceptional work; proper assessment of opportunities of [sic] expansion, development and relationships that align with our current position and future endeavors; intentional travel with specific goals and appropriate budget in relation to our expansion, development and relationships; transparent bookkeeping records and democratic budgeting, affording the ability to maximize monetary resources across all sectors of OSBG.[10]

The letter claimed staff would halt all operations immediately and would resign unless leadership responded. In the end, nearly the entire staff and most of the students left the organization. Some of them remained in the city and moved onto other existing urban cultivation projects or started their own. Our School at Blair Grocery went on hiatus for a few months, then regrouped and resumed operations. The leaders viewed the incident as a moment to "reset." When I asked them about it a year after the walkout, they indicated that they had downsized their operations and began investing in other priorities. The organization would continue to operate for seven more years before the two leaders left the city and the project altogether.[11]

The very public nature of the OSBG fallout was unique and specific to a project that dealt with other internal challenges.[12] The organization's financial management also faced critique from other growers, especially because OSBG had been very successful in winning large grants. More broadly, though, the critique reflected the troubling patterns among many nonprofit organizations that operated in the Lower Ninth Ward, and received a level of funding and media attention that did not necessarily match community outputs, particularly in the eyes of local observers. Using a community's trauma to promote the work of charitable or "social justice" organizations unfortunately seemed to resonate well with foundations eager to support these causes in New Orleans, contributing to an exploitative affect economy.[13] The OSBG walkout incident demonstrated how a project that started with a relatively radical and prefigurative aspiration to create an alternative educational space for young people could get caught up in its eagerness to experiment and grow quickly without the range of skillsets needed to retain focus and consistency, resulting in organizational disruption, negative impacts on the participants, or worse.[14]

Limited Skill and Capacity

The growers who came to urban cultivation with limited horticultural experience and knowledge found it a challenge to scale up their projects from private gardening or community gardening. Skill limitations were most common among those who entered urban cultivation serendipitously during the transitional period, but even the experienced growers had to learn to adapt to new climate and soil conditions if their previous cultivation knowledge came from other parts of the country.

Colleen described a lack of horticultural skills as the major hurdle for sustaining cultivation projects:

> Skill, yeah. Skill is a huge issue. It takes a lot of time to manage these things. And when you get something like rooting out nematode and people with the big these letters behind their name [that indicates advanced degrees] can't even help you . . . I do [grow organically]. But there's fertility issues that people aren't really being able to address. There're disease issues that people aren't really being able to address. So I could spend half of my year trying to figure out how to get the nematodes under control, and (then) get a totally different property and have a different problem.

The knowledge obtained in classrooms or even workshops could only take growers so far. Nor were growers' interactions with experts with "letters behind their name" especially helpful, particularly if those experts specialized in rural agriculture. One grower even went so far as to claim that the "master gardener" course he took was "useless" when it came to applying skills at the scale of cultivation he was trying to implement. Each season brought new challenges to all growers, which meant that learning and experimentation never ended.

Horticultural skills were only one part of the skillset growers needed to run their projects. Depending on the type of cultivation projects they were operating, growers also had to engage in community outreach, fundraising, networking, marketing, accounting, and advocacy. Growers with various professional backgrounds in teaching, law enforcement, nonprofit organization, and entrepreneurship had some previous exposure to leadership roles or management experiences. But no single grower entered urban cultivation with all of these skillsets. Some collaborated with others to complement their skillsets and capacities, while others developed needed skillsets on the job.

Most of the urban cultivation projects were one- or two-person operations. Only a handful of organizations became large and structured enough to have full- or part-time staff to take on specific tasks, but, as we have seen, organizational complexity required managerial skills that growers didn't necessarily have. In small operations, many responsibilities often got overlooked, under-prioritized, or unfulfilled as the growers stayed focused on their priority—growing. When growers' inability to keep up with these competing demands became overwhelming, or when the accumulation of failures caught up with them in the forms of audits, regulatory requirements, or disgruntled collaborative partners, it threatened the project's sustainability, as the walkout at Our School at Blair Grocery exemplified.

Jenga reflected on what she learned about how to build capacity, both financially and socially, by recognizing the limitation of doing it all on her own:

> I think I looked at that, in a very noble kind of way, "we're doing all this stuff on a dime," like "We don't have anything. And yet we're doing all of this!," you know? And part of that is beautiful. At the same time, it's not sustainable. Like, you cannot set the expectation that you will always do more than [what] is expected with nothing. . . . But operating at capacity is "okay, it's actually only reasonable for a human being to spend, you know *x* amount of time doing this task. And to be compensated for this task with a living wage or something that's comparable to the work that they're putting out," and that's it. And then if you need more done, then you need to bring on another person. And so really building it out and making sure there's money there to pay people.

The implications of Jenga's realization became especially acute when she started her graduate training and could no longer engage in the day-to-day operations of the two gardens and community outreach in the way she used to be able to do.

Practitioners of prefigurative urbanism frequently encountered roadblocks and had to take on a multitude of roles, many of them unanticipated. The desire to manifest something that does not yet exist means constantly adapting while moving forward. In some ways, over-commitment felt like a validation of their work as significant and mean-ingful. Yet, as Jenga noted, these additional tasks and responsibilities build over time, making it difficult to recognize when they had reached the capacity threshold.

Life Stages

Moments of reflection could lead the growers to assess their engage-ment with prefigurative urbanism. The growers who turned to urban cultivation as a form of personal inspiration and fulfillment were often on a journey of their own, as described in Chapter Two. Members of this category of grower often began their projects at a transitional period in their lives, even if they didn't realize it at the time: a person's late twenties and early thirties are often full of significant life-stage changes in their relationships, career, and financial situations. Just as most social movement activists' engagements ebb and flow across their lifetime,[15] these growers found that what worked as a life routine at the beginning of the project did not always work for them as they got older.

Caroline encountered this revelation somewhat differently than most of the growers, having pursued urban cultivation after a long-term ca-reer in the film industry. When I asked her about the final straw that led her to shut down her entrepreneurial cultivation project after "just under three years," she responded:

> I was just sort of getting to the point where I understood what was not worth doing anymore or not trying to do in that level of scale. I just hit my fiftieth birthday, and it was pretty much like, "Okay, I've made a lot of money in my lifetime and I've made very little. This is a time of my life where I need to not be making very little. I want to work for another ten or fifteen years and then not work anymore. If I continue to pursue this life, I'll be scraping for as long as it lasts." It was going to be a lifetime of poverty, because I'm trying to live in a city and have city expenses and survive on a farmer's income. There was a disconnect there. I had very low overhead. Even with that, I couldn't make sense out of it.

Someone who is younger or at a different stage of life might have been willing to take on this level of financial insecurity, but people's needs and their life circumstances change, as Caroline's comment implies. She described her experience as: "It was fun. I felt like I was doing something cutting-edge and different, and people, the public was really into it." But realizing the limitations of pursuing a career as an urban grower, combined with the insecurity of her land access and her outlook on later life prospects, she "couldn't make sense out of" being a full-time urban grower anymore. For some of the older growers, the physically demanding nature of gardening and farming also raised questions about how much longer they could continue the work at the same pace and scale.

The impact of life stages on growers' decisions to exit urban cultivation must also be contextualized through the lens of intersectionality, with attention to race, class, and growers' relationship to the city. The growers who saw their engagement with urban cultivation as a personal endeavor tended to be white and middle-class, or did not have familial connections to New Orleans. The individualistic nature of their decision to leave their cultivation projects reflected how they got into them in the first place. But who *gets to* have opportunities, or at least feel entitled, to think of these life decisions as personal ones?

While discussing the variant trajectories of the post-Katrina transplant growers in 2014, Cory wondered out loud if many of them were in "a transient moment in their lives." Many moved to New Orleans and started the cultivation projects to tend to the vacant land "in the way that makes sense to them":

> It's not just about access to land, but what is it about the sort of privilege of mobility, and being able to start projects, you know? In 2012 we saw an increase of newcomers moving to New Orleans, in a context, however, of Black New Orleanians still not really being able to come back post-Katrina, you know? And so what are the sort of privileges and support at play that made it more possible for white Northern transplants to start a [cultivation] project?

Cory's questions highlight how some individuals' decision to engage in prefigurative urbanism reflected their class and social privileges, from race and class to career history, that afforded them with a sense of individual choices and ability to act on these aspirations. Their transitional place in life could have heightened their need to "make sense" of their actions, becoming an impetus for their engagement with prefigurative urbanism. Urban cultivation was a way for them to enact a sense of hope about their own future, not just altruism for making change.

To be clear, engaging in prefigurative urbanism does not by definition require access to privilege and resources, though financial stability does increase growers' capacity to sustain their operations, as I have established above. As we saw in Chapters Three and Four, privilege did not guarantee the successful launch or implementation of a new urban cultivation project, let alone land acquisition. When a project failed to sustain itself, or a grower decided to depart urban cultivation or New Orleans, it did not necessarily mean that the individual involved did not value the work or the people their work intended to support or empower. But as a form of social change that does not start with collective organizing or mobilization, prefigurative urbanism starts and ends in these form of individualized projects unless participants actively seek ways to develop collective ownership or successors to sustain the project past a founder's tenure.

Burnout

Both internal and external factors may lead growers to terminate a project or leave urban cultivation, even if they would have preferred to continue. But there were cases when growers felt they were done with urban cultivation altogether: when they experienced "burnout." Social movement scholarship points to three sources of activist burnout, including motivational and emotional factors, organizational culture, and within-movement discord.[16] Activists from marginalized, frontline communities face an additional risk of experiencing burnout in the form of their critical and constant awareness of the scale of systemic social injustices and their self-imposed sense of responsibility for the community they represent.[17] Experientially, burnout is not merely exhaustion or loss of motivation, but a sense that an actor no longer views the return on their work as worthy of their efforts. In other words, burnout indicates a sense of hopelessness in one's endeavor.[18] Absent the community so crucial for social movements, prefigurative urbanism loses its purpose when the actor loses this sense of willful hope, whether for themselves or for others, underscoring McGeer's argument for "responsive hope" to avoid such pitfalls.[19]

Burnout mostly happened to some of the younger growers who entered urban cultivation in their late twenties or early thirties, as well as those growers who had managed to sustain their projects for at least five years. In the case of urban cultivation in post-Katrina New Orleans, we might expect the physically demanding aspects of urban cultivation, especially in the hot and humid conditions of the city, to be one of the major drivers of burnout. But for most of the growers, regardless of their age, the sense of burnout did not derive from the physical work itself. When I asked Caroline what part of the work she found most challenging, she responded, "It wasn't being out in the hot Louisiana sun. It was none of that. All of that was fine." She shared that she had "learned to live with it and it was fine," and reflected that she enjoyed being outside, even in the heat.

Strenuousness of the work itself, then, did not discourage the growers. After all, the growers partly pursued urban cultivation because they wanted to be challenged by the physical and tangible aspects of the work. It was *everything else* that contributed to burnout. Erin observed that the burnout she saw among the growers around her had more to do with the nonprofit aspects of the work than the farming itself:

> I think there is this phenomenon of the nonprofit burnouts. I've talked to people about that. Kind of the cycle of burnout that happens in the nonprofit world. I don't think farming is necessarily [the cause of burnout], if all you're doing is farming work. I think urban ag is inherently more community-based, because you're in the city. So it looks a little bit different from [rural] farming. I do think that nonprofit burnout in combination with urban agriculture is different.

All growers acknowledged that maintaining urban cultivation sites was hard work. But the sense of burnout stemmed from their being overwhelmed with the non-cultivation aspects of the work, a situation that was exacerbated when they lacked financial security, the breadth of necessary skills, and organizational scaffolding, as described above. Mission creep also threatened growers' morale—when they felt like they were no longer doing what they had initially set out to do.

In social movements, activist burnout has often been linked to problems of organizational culture.[20] But these sorts of issues, such as infighting and internal marginalization, tended not to come up as often in urban cultivation, presumably because most growers had not embedded themselves in organizations. Instead, growers' overemphasis on direct action and the individualized nature of their practice became causes of burnout distinct to the urban growers engaging in prefigurative urbanism. The ends-effacing nature of prefigurative urbanism meant that assessments of whether they were headed "in the right direction" remained irrelevant, elusive, or at least not the priority for most urban cultivation projects during the initial years of formation and development. As the projects scaled up or expanded in scope, the existing challenges in land and financial insecurity, limited resources and skillsets, and the need to constantly adapt and experiment all began to take their toll on the growers, and they found it difficult to assess if it was all "worth it." The inherently individualistic nature of their praxis created an additional challenge in implementing the kinds of collective, preventative measures that social movement organizations can pursue to avoid burnout. It is noteworthy that it was not uncommon for growers to remark during our interviews that they appreciated the moment to reflect on and take stock of their work, something they rarely took the time to do on their own, though in other cases a few noted that follow-up conversation was not something they looked forward to because they feared it could be triggering of the emotional distress they felt when they decided to leave the practice. These comments further underscore how much of the growers' energy was focused on present moments as they strove to continue prefiguring their visions of alternative future.

Cultivating Persistence

In the same way that no one factor was sufficient to guarantee a project's demise, no one factor would ensure an urban cultivation project's sustainability. Land access security was crucial for the long-term viability of a project, but the struggles and eventual termination of Our School at Blair Grocery remind us that land access alone does not guarantee the longevity of an urban cultivation project. At one point, OSBG had one of the largest cultivation spaces in New Orleans; the organization had been successful at securing large foundation grants for nearly a decade. It even survived, in a new form, after the staff walkout. Nevertheless, its leaders ultimately quit urban cultivation, citing burnout. Unmentioned, but likely contributing to the end of the project, were other challenges that had accumulated over time, including poor financial management and mixed results in community outreach.

Community buy-in does increase the likelihood that a cultivation project will endure, but only to a degree. When Cory decided they needed to step back from their role in leading Good Food Community Farm because of personal health challenges, a community member who had been working with them stepped up and took over the operation:

> Ms. Gloria started taking over stewardship of the land. Not alone. There [were other] long-term community gardeners still involved. But Gloria really took on the helm of like, stewarding the land as a social project in the neighborhood. She's like the kind of person who galvanizes resources and just, like, excites people, so she was doing children's programming and made this beautiful fence on Claiborne with a bunch of big tires, painted in the yellow bricks, had a literal yellow brick road, had a stage with a piano. Like really actually took the project, the space, in a direction that was so beautiful. And was so brimming with potential. It became the space that hosted bi-monthly, or like a couple of times a month, farmers markets, where farmers and Black business people could vend. [The space] regularly hosted concerts and really just kind of moved the space in a new direction that I just didn't, you know?

The practice's transition from the leadership of a white, post-Katrina transplant to a long-term Black community member demonstrates the potential of community engagement to create a pathway for continuity and community ownership of a project originally started by newcomers and outsiders.

Not everything about Cory's original project continued through the transition. The market struggled to sustain its sales of microgreens and vegetables. Cory thought this was fine, noting, "Everything has its time." The garden also changed its name, which did not bother Cory

Figure 5.3. One of the former Good Food Community Farm spaces in Tremé neighborhood, 2023. Photo by author.

either. But then came the land loss. One by one, the lots they had been operating on got sold, and the growers were kicked off the land. The razed lots sat empty for years, presumably as the new owners waited for the property value to appreciate (see one of the lots in Figure 5.3). Gloria moved to another, much smaller, space nearby, where she resumed offering community programming, albeit at a reduced scale. Having the trust and commitment from community members facilitated Good Food Community Farm's initial transition and provided resilience for the project to continue the practice in the face of eviction, but the forces of gentrification do not defer to how much a community values a project.

Though the growers worked independently, they supported each other in resolving problems from time to time. In fact, they were more likely to credit informal mentorship and support among the local growers' network over formal training, like the Master Gardeners programs, for their capacity to succeed. On dozens of occasions, I observed one grower crediting another for an aspect of their growing practices, or growers referring to each other for advice, even as they disagreed on specific philosophies or "best practices." Eric described his relationship to other growers he knew in the city as a mutually supportive learning system:

> We just talk to each other, we learn from each other, some things like that. So I think the majority of people [do this], you know, but there are some people who firmly believe that their way is the only [way]. And that's fine, they get to. But I think a lot of it is that people just wanna learn, and they learn by trial and error. I mean, everybody goes into this, they got different locations. I mean, you're talking, if you put out a rain gauge between maybe a mile apart from here, you'd get a totally different range. . . . It's a different environment. So what they do is they struggle with their struggles and that's how I know about them. They call me up or I used to deliver soil to a lot of growers, and you know, when I talk to them, "Hey, how are things going?" First things out of their mouth is "It's really good" or "[It's] really bad." [I'd then ask,] "Why?" Boom, and they'd tell you. And if they have a bad problem, I'd look into it for them. When one of them had a leaf freeze or something like that, the guy said, "I'm gonna go buy some nematodes." Oh cool. So I'd buy nematodes. Issue went away. We're all just kinda learning.

While their approach valued individualized action, the growers constituted a loosely connected network in New Orleans, with some of them regularly communicating with and offering support to one another in various forms. There was nevertheless a tacit understanding that everyone was implementing their own ideas and should ultimately be left alone, reflecting the individualized nature of prefigurative urbanism praxis. The growers' ability to capitalize on their existing human and social capital was typically not sufficient to overcome blunt external forces, such as loss of access to land or an accumulation of internal factors. Nevertheless, those with networked support could reduce the frequency and scale of individual challenges, which made those challenges more manageable and created a sense of belonging, even if it was ephemeral.

Those cultivation projects that did not expand too quickly beyond their capacity, or that approached new opportunities with caution and pragmatism, more often avoided abrupt setbacks from land loss, market failure, internal organizational discord, or burnout. When we originally interviewed Jenga in 2014, she described her future plans expansively, but with restraint:

> Right now, we really want to just like perfect what we are doing now. And we really want the Guerrilla Garden, especially, to be a model on how, you know, gardens can be true community centers. When a lot of people hear about a garden, they automatically think, "Well, you have to be a gardener if you want to be involved," and "You have to like gardening," and "You have to like what it's all about." And [they assume] it's about producing food. And it's open for us. We want to show that it's also open for everybody to come. Everybody likes to eat. I hope everybody likes to eat, you know. [Laughter] So there are so many different things that you can do, you know, with a garden. We really want to have the Guerrilla Garden be a model for that. As far as growth, we don't have any immediate plans for creating other gardens or making our gardens bigger than [what] we have right now. We really want to focus on just perfecting what they currently are.

During our follow-up conversation in 2020, Jenga revealed that the garden had not acquired any more cultivation spaces and remained focused on providing gardening opportunities and programming for community members. But her reduced capacity, the result of her attending graduate school, in addition to the loss of access to the house adjacent to one of the gardens, led the organization to reassess its organizational resources through a period of strategic planning. Plans emerged for the garden to collaborate with a few Black women in the community who were doing work around plant medicine and in supporting Black mothers, with the space to be used for their programming. By pivoting to identify collaborative opportunities and staying focused on long-term Black New Orleanians, Jenga avoided a situation in which her personal life changes and looming burnout led to the termination of the project. Other growers who recognized their limitations would simply return the leased lots when they decided the space no longer suited their needs or capacity.

Longer-term goals for the garden sites, including planting low-maintenance plants, such as fruit trees, composting, and incorporating beekeeping, allowed the growers to adjust the day-to-day need for maintenance. Zach explained that initially he did minimal work on the lot he leased from NORA, because he was not sure if the agency would actually sell him the lot at the end of the two-year lease. To his surprise, NORA agreed to go through with the sale in 2018. But he nonetheless decided to take a long-term approach to developing a cultivation project in the space, including planning for his own retirement:

> My intention is to plant some trees around the edge, grow some peren-
> nial things. And then at some point in five or ten years, that might be my
> retirement. Put a little home on there, you know? Rent that whole house,
> sell it, something like that, but keep it where it at least starts and contin-
> ues where there's permanent stuff that will be there, regardless of whether
> the house is built.

In keeping with his visions of having food-producing plants that are low-maintenance for as many people in the community as possible, his plans foresaw how the space could continue even in his absence, or when he no longer could or wished to engage in day-to-day cultivation work. Nevertheless, reflecting the ends-effacing nature of prefigurative urbanism, he did not feel it necessary for him to decide what, specifi-cally, that project would look like. He just wanted to create opportunities for others.

Sustaining Prefigurative Urbanism

As this chapter has shown, a cultivation project that has been terminated did not necessarily "fail." The cultivation space can continue to thrive as a productive garden with a new grower, or the grower may move to another place or practice to continue prefiguring the future beyond their original project. When I asked Pamela to reflect on one of the projects that she had to discontinue due to safety concerns, she responded that she saw it as a success because of what it *did* accomplish, even if it did not have the longevity she had originally intended:

> A lot of people are saying, "Ok, so if I invest all of this money on the front
> end to put in infrastructure on property that I don't own, that maybe I
> might be able to purchase after three years; suppose that falls through,

then what?" For me, when I was at Sun Harvest, I knew I was going into it as a risk. But I was willing to take that on, because my primary intent was demonstration, you know, like how to properly set it up. So even though I had to take the garden down because of the reasons I had to, for me it was successful. Totally successful. I'd say that I was able to see the good, the bad, and the ugly of urban growing, so I didn't fall apart when I had to take it out. I said, "Ok, so if this occurs what is the best thing that I can do?" And for me, it was to share the fencing, soil, and plants with other growers to help them build greater capacity. I knew that I would be alright, you know. I didn't say to myself, "Oh my God, my business has failed." I wasn't in it for that.

Pamela's insistence that Sun Harvest Kitchen Garden was a "success" was not a defensive positioning of the project she worked hard for several years to launch. Pamela viewed the project as a success because she manifested a garden on a formerly blighted property, convinced a nonprofit organization that such a garden was worth its commitment, and showed the community of nearby residents and fellow growers how to set up an ecologically sustainable and productive garden. She furthermore anticipated at least some of the challenges she encountered, partly because she had been gardening in New Orleans for most of her life and had seen gardens come and go, but also because she was aware that what she was attempting was new and had no institutional, legal, or social guarantee. When Pamela determined that the garden could not continue given the increasing presence of an unhoused population, she did not fight to keep it open. She understood her specific challenges as resulting from a complex set of political, social, and economic factors, and she did not expect the city to respond to the issue in a manner that would make a difference for her garden. What she did, instead, was to pass on the resources, both physical and informational, from this project to other gardeners, and then she moved on to other urban cultivation work.

Planning for long-term sustainability beyond the founding individual's tenure proved difficult even when growers knew the importance of planning for transitions. There were a few exceptions. For instance, Grow Dat Youth Farm continued and even expanded with a new, more horizontal leadership structure after the founder departed. Its organizational resilience was tested in 2024, when the City Park Conservancy announced a plan to put a new road through the farm. Grow Dat Youth Farm was initially blindsided by this plan, as the organization was never consulted prior to its public release, and such a disruption of the space would have virtually disabled its operation. But it quickly organized itself and successfully rallied strong public objection to this plan, led in part by current and former youth participants, and eventually secured a longer-term lease with more transparency with the City Park Conservancy.

In most cases, however, it was not easy for founding growers to find someone who not only shared their ongoing vision of the project but was also willing to commit for a long time. In part, this situation was a consequence of prefigurative urbanism's emergence as an individualized set of actions with evolving aspirations and praxis. Most cultivation projects in New Orleans were run by only one or two full-time growers, and interns and apprentices only engaged with a given project for a short time, often with the expectation that they would move on to start their own project rather than join or inherit the original project down the road.

Continuation of work also means supporting other growers, even if your own project may terminate. A grower in 2018 observed that there had been people who came to post-Katrina New Orleans because they heard that it was like the "Wild West," with little regulation, but for one reason or another they did not have what it took to persist. This grower observed that the practice of these newcomers contrasted with that of long-term growers like Jeanette and Pamela, who had been pushing for the kind of infrastructural and public support for urban cultivation for a long time with limited success, especially when it came to policy advocacy. She saw that there was another group of younger people that came "bright-eyed and bushy-tailed and naïve," counting herself among this group, who are trying to carry on and further advance Jeanette and Pamela's efforts:

> I think what we're trying to do at this exact moment is to organize and to be more critical in the way that we spread information to share resources to be more collective with new and younger growers, rather than competitive. We really try to *cooperatize*—made-up word—and rethink what it means to do urban agriculture in New Orleans, because it is such a

specific place. . . . There are young people [here] who are tired, but we're still [asking ourselves], "How do we reimagine this, like, how do we do this right?" I think that we kinda feel like we are at a point where a lot of us are like, "We have to leave or give up," and we aren't ready to give up.

These observations highlight that growers, while operating individually, do think about the collective impact of their actions in aggregate, especially in the long-term. Keeping the work going did not have to happen on the same lot or by the same grower, but a project that functioned as a catalyst to other projects could also be considered a form of success, since it contributed to the larger and sustained manifestation of urban cultivation practice in the city.

Transitioning beyond Prefigurative Urbanism

A handful of urban cultivation projects transitioned from prefigurative urbanism to something more closely resembling a social movement. There were always some efforts to organize urban growers over the decade. Around 2008, a group of urban cultivation projects formed The Lower Ninth Ward Urban Farming Coalition. The website (no longer active) described it as:

> a partnership of local groups and individuals committed to food security and environmental responsibility through urban agriculture. We are working to enable food justice through establishing a sustainable food system in the Lower Ninth Ward, a neighborhood in which there are no grocery stores and limited healthy food options. We are gardeners, farmers, nonprofits and dreamers who want to grow fruits and vegetables grown in the Lower Ninth Ward. We help each other out, and try to get our hands dirty as much as possible.[21]

It listed four projects that were active in the neighborhood at the time as coalition members: the Backyard Gardener's Network, the School at Blair Grocery Micro Farm and Youth Stewardship Program, Common Ground Relief Urban Agriculture Projects, and the Villere Street Farm. The coalition was loosely organized, without specific leadership, and seems to have become relatively inactive by 2010, just as each of the four projects began adapting to new opportunities and interests of their own. Recall from Chapter Four that Cory, too, organized dinner conversations that lost steam after about a year. While these efforts did not continue in the long term or give rise to a collective movement, they connected growers across the city who continued to give support to each other individually.

In fact, Greater New Orleans Growers Alliance (GNOGA) persisted over the years, with members meeting regularly to discuss a range of concerns and giving support where needed. But membership in the organization has been selective and not representative of the full spectrum of urban growers in New Orleans. Margee, who had been actively engaged with the group, shared similar observations in separate conversations about why it took the trajectory and form that it did: Participation by growers fluctuated, which created varying interests and concerns among the members. Conversations mostly focused on solving immediate issues, such as water access, land tenure, or pest remediation, rather than addressing policies or broader food systems. Many growers rarely or never participated in the GNOGA, for various reasons. But Shannon, who had also been active in the group, saw this through a more nuanced lens, looking beyond an evaluation of success or a failure. While she wished that GNOGA had been able to organize more strongly as a collective, she also recognized that urban growers are independent and pragmatic actors. Thus, she saw the current stage of the organization—a loose network of individuals who came together as needed to take on concrete issues, rather than forming a social movement organization with a formal leadership, membership structure, and shared collective goal—as a step toward a long-term trajectory of building power.

Conclusion

Urban cultivation projects that began as a form of prefigurative urbanism may end in many ways, for various reasons, but never for just one reason. We may expect projects with deeper connections to the local community to be more successful in sustaining their practice, compared to the projects that were initiated primarily by grower aspirations for an alternative career or lifestyle. Yet the fate of any urban cultivation project in New Orleans was unpredictable, in part because the projects continued to evolve past their original aspirations and practice. Urban cultivation projects were operating in a city that was undergoing rapid social and economic transformation. With their focus on immediate, tangible, direct action, some of the projects were not well positioned to face these changes. Urban cultivation projects that operated as a form of prefigurative urbanism set out with neither a specific goal nor a planned end point. This complicates how we define and assess the termination of a prefigurative urbanism project; by design, wherever it ends is how far it was going to get.

The end of an individual project does not signify its failure or insignificance. Even when it does not produce a long-term, sustained practice, prefigurative urbanism creates a "wake" in its path. It can give birth to another project, or the actor can continue prefiguring alternative futures in another realm. Theoretically, it could mobilize a social movement or evolve into a prefigurative political practice. At the least, the actors involved demonstrated to skeptics that alternative ways of living, working, and using the land in the city were possible. In this way, prefigurative urbanism can be generative beyond any given project or actor. Most importantly, however, the lack of predetermined outcomes creates open-ended trajectories that encourage persistent imagination, innovation, and exploration, a mindset that prioritizes action over ideals and values. This is particularly important in transitional urban contexts. One must be on guard and keep moving forward to forge ahead into the future.

Unfortunately, the ripple effects of prefigurative urbanism are not all positive. If one grower's intent was to convert blighted and undervalued spaces into green, biologically and socially productive spaces, then letting the space return to blight could signal abandonment not just of the project but of the hopes that the space intended to cultivate. The fact that growers could leave to pursue other prefigurative urbanism practices or discontinue their engagement with the practice altogether underscores the privileges and inequities that restrict who is able to take on this form of civic engagement in the first place. The financial and other resources needed to start and sustain prefigurative urbanism projects exclude those with limited resources, including time and cultural capital. Here, racial and class privileges that afford white, middle-class actors more opportunities and a sense of entitlement to launch prefigurative urbanism projects must be critically recognized, especially when these practices are implemented in communities of color or in low-income neighborhoods as I elaborate in the next chapter.

Participants in prefigurative urbanism opt to take individual, direct action to implement social changes, which reflects both their strong sense of self-efficacy and their distrust of existing social and legal systems. As a result, the community of growers interprets the termination of a particular project or a grower's departure from the city as an individual decision rather than a collective loss or failure. To be clear, leaving is never easy, and those who left urban cultivation or the city altogether often expressed a sense of loss and regret. Some of them later explained that they felt a need to distance themselves from the experience in order to physically and emotionally heal. Yet these growers' ability to depart, temporarily or permanently, also distinguishes them from those who remain in these communities, with or without choice. The prefigurative urbanism practitioners who chose to stay and continue their practice, therefore, were no doubt strongly committed to and skilled at growing and adapting their projects. But they also were able to access the financial and social resources to continue their practice while maintaining their willful hope.

6

Cultivating Hope in a Post-Disaster City

Immediate Changes, Variable Outcomes, and Uncertain Futures

The post-disaster urban cultivation practices I have discussed in this book do not fit squarely into the existing narratives about what happened in post-Katrina New Orleans, most of which focus either on neoliberal redevelopment that exacerbated inequality or grassroots responses that called out social injustices.[1] Urban growers' actions primarily took the form of prefigurative urbanism and interacted with both of these competing forces—neoliberal redevelopment *and* grassroots responses—while charting their own course. This chapter explores what the variant aspirations and trajectories of the urban cultivation scene in New Orleans reveal about the potential and limits of prefigurative urbanism in a city undergoing social, economic, and political transformation. I draw on theories of urban development, disaster recovery, social movements, and environmental and food justice to examine why some people choose to engage in this form of civic engagement over more conventional political activism, and what the significance of these actions in a post-disaster city is.

Several exogenous and endogenous factors contributed to the emergence of and divergent trajectories within urban cultivation in post-Katrina New Orleans. Institutional failures to prepare for and respond to the disaster reinforced public distrust of the government's ability to address social issues. Nonprofit organizations and grassroots activists became the dominant forces to provide resources and advocate for the residents who could not evacuate, return, or rebuild, but these efforts presented their own limitations and challenges in the face of a vast and multifaceted structural catastrophe that reflected and exacerbated pre-disaster conditions of social stratification and injustice. In this context, prefigurative urbanism's emphasis on individualized, immediate direct action resonated with some individuals as a way of implementing social change. Many were experiencing transitional moments in their own lives and had the dispositions or privileges that made them able to do something that did not fit the existing social and legal norms. Overall, prefigurative urbanism produces variable concrete, short-term outcomes as the actors continue to experiment and adapt, but these actions also have unintended and unpredictable long-term impacts.

Realms and Modes of Social Action in Times of Uncertainty

Individuals, groups, and organizations may take several different approaches to social action, that is, in how they attempt to effect change in society beyond their own personal interests and benefits (Table 6.1). On one end of the spectrum are people and organizations that refrain from participating in social action altogether. Their inaction could be due to a loss of hope in potential resolutions, ignorance, incompetence, or a lack of capacity or sense of efficacy, any one of which could prevent them from taking action. Their willful or unintentional disengagement perpetuates social problems, which they then read as further evidence that there was no point in trying to make changes. Scholars have developed several theories to explain declining civic engagement and the associated "free-rider" problem, when people who would otherwise benefit from the changes that activists are trying to bring about do not join a movement. Low voting rates and other forms of declining civic engagement have been attributed to a range of societal factors, from competing demands on time to technological advances, demographic changes, and general loss of social capital.[2] At the other end of the spectrum are individuals actively engaging in addressing social problems. The space between the two ends of the spectrum, where individuals take some form of action but do not become explicitly political or join collective organized actions, has been largely overlooked in studies of political sociology and social movements. Who are the individuals doing *something* about social issues, but not as activists, and why would they choose this particular form of social action?

On paper, voting may be the most accessible means of demanding social change, yet voter suppression and limited candidate options make it difficult for many voters to feel confident that their vote matters, particularly for those from marginalized backgrounds, though

TABLE 6.1. Forms of Social Action by their Modes of Action and Constraints

Realm of Social Action	Mode of action for social change	Constraints
Government	Policy and programs, Funding	Lack of transparency, Bureaucracy, Political compromises, Slow pace of change, Lack of resources
Civic engagement	Voting, Volunteering	Voter suppression, Slow pace and compromises of the legislative process, Limited time and energy for volunteering
Market	New products and services, Corporate Social Responsibility, Selective consumption of socially conscious products or businesses	Prioritizes economic return, Exacerbates inequality, Class and status limitations in who can participate, Potential to be co-opted by corporate marketing
Non-Government/Non-profit Organizations	Improves access to resources, Policy changes, Fundraising	Nonprofit-industrial complex, Mission-praxis misalignment, Funding instability
Social movements	Organizing, Demand policy changes, Improve access to resources, Direct action	Time and compromise required for consensus-based collective approach, Slow pace of change, Free-rider challenges
Prefigurative politics	Direct action in public realms, Enacting the future outside of the status quo	Regulatory sanction limits transformative potential, Risks of co-optation
DIY urbanism, etc.	Direct action through public spatial modification and deviant spatial use	Ephemeral and partial modification, Art and design-focused actions, Privilege of actors
Prefigurative urbanism	Direct action in public realms, Enacting alternative futures	Changes limited to hyperlocal solutions, Open-ended visions and trajectories, Mission creep
No action	Apathy, Ignorance, Incompetence, Lack of capacity or efficacy	Problem remains or worsens, Contributes to self-fulfilling prophecy

also increasingly among conservative voters.[3] Even when individuals may wish to vote, their ability to participate in the electoral process is limited by laws that strip them of their rights, by gerrymandering and voter suppression, or because of citizenship status. More recently, social movement scholars have begun to theorize individuals' non-participation in social action as a rational response that results from a lack of trust in the system's ability to change.[4] In fact, social action that takes the form of direct policy changes, whether at the local, state, or federal level, may be the most impactful legally, but the legislative process is slow and involves compromise and bureaucratic red tape. The agencies tasked with implementing these policy changes are frequently under-resourced, politically constrained, or infiltrated by industry lobbyists, rendering them ineffective with various loopholes that fail to address underlying issues. It is not surprising, then, that public trust in government's capacity to enact social change has been steadily on the decline in the US since the 1960s.[5]

Market-based solutions to social issues have meanwhile gained popularity among policymakers, entrepreneurs, and investors. Adherents to this neoliberal perspective believe that market competition will produce the best outcomes, even in the realm of social policy; they trumpet the virtue of individual responsibility over state mandates and government assistance. Since around the end of the twentieth century, the notion of corporate social responsibility (CSR), which promises a positive impact through "sustainable" practices or large-scale donation, has gained traction with investors. Consumers, too, have been encouraged to partake in market-based solutions by participating in so-called "lifestyle movements."[6] Purchasing organic produce at farmers markets or patronizing small, locally owned shops have become most prominent examples of conspicuous "ethical consumption" practices,[7] in which the act of consuming comes to be equated with civic engagement, through expressions like "voting with one's fork." This market-based approach to democratic participation requires economic means and cultural capital, even as the consumers engaged in this mode of social change tend to underplay the privileges and status that enable this form of social action.[8] Academic research and activists have documented the unequal distribution of the "benefits" that supposedly result from these market-based solutions, pointing to the weakening of the public social welfare system through privatization that puts vulnerable populations at risk and exacerbates the very problems that corporations claim to be solving.[9]

In situations where the government has failed to effectively address social issues, nonprofit organizations step in to provide resources, services, and advocacy. Since the late 1980s, the rollback of public safety-net programs in the United States has steadily increased the need for these non-government entities, creating what activists and scholars call the "nonprofit industrial complex." Instead of challenging the existing policies and economic systems that create social disparities, critics argue, these organizations have developed symbiotic relationships with government agencies and philanthropic foundations that justify their own existence, thus strategically avoiding a situation in which they have worked themselves out of their jobs by addressing the roots of the issue.[10] Nonprofit organizations rely on both public and private funding for their operations, creating competition for and uncertainty about the sustainability of their work. The competition for funding and media attention sometimes produces mission-praxis misalignment, in which organizations redesign their services to appeal to key donors or supporters. Many large nonprofits additionally feature a hierarchical, professional structure that more closely resembles corporate organization than a grassroots movement.[11] Thus, despite the prevalence of nonprofit organizations, engaging in social action through them comes with organizational, financial, and bureaucratic constraints not too dissimilar to those of governments and corporations.

Many nonprofit organizations come into being when grassroots social movements decide to formalize their operations. Individuals who desire to address a problem they see in their community or in society at large can form a group based on shared grievances, or they can join an existing group or organization. While this form of collective organizing can be very effective, it takes time to connect with, educate, and recruit members, articulate a shared grievance and mission, agree on a strategy, and develop effective actions and community outreach. Scholars have variously understood people's hesitation to join movements as stemming from a lack of incentive;[12] exposure to risk, particularly in hostile environments;[13] and a lack of economic or social capital to afford time for participation, even if they cared and wanted to join. It may be a long time before the outcomes of these efforts materialize as changes in policy or social norms. Actors must persist in their efforts across ebbs and flows of the social and political climate over time, often having to wait for the right moment when their issue gains saliency or, alternatively, having to adjust their messaging to identify new supporters.[14] Moreover, working on social issues through these realms still requires a level of faith in the possibility of institutional change, such as believing that the policymakers can be persuaded or that policy change matters. It requires patience, because societal-level changes are often incremental and not directly attributable to singular, organized efforts.

Those who do not trust the status quo to relinquish its power engage in prefigurative politics. In prefigurative politics, activists prefigure a future that is yet to come, such as social revolution, racial equity, or the dismantling of capitalism. Their direct actions are motivated by intentions of addressing issues of "power, conflict, and transformation."[15] By bypassing the process of planning and waiting for change, prefigurative politics challenge the distinction between the ends and the means, or the future and now, of a movement.[16] The potential of prefigurative politics to scale up or fundamentally challenge the existing social, political, or economic regime is uncertain and depends on external circumstances.[17]

DIY (do-it-yourself) urbanism is a concept from urban design scholarship that describes situations where "non-professional or unauthorized community members create or modify urban spaces in ways typically considered to be the purview of municipal employees."[18] Similarly to prefigurative politics, DIY urbanism is a form of political direct action that aims to undermine the status quo, specifically in urban spatial design, land usage, and management. Many similar concepts describe various forms of temporary physical alteration or occupation of public spaces that challenge the legal and social powers of the government and professional urban planners and designers: tactical urbanism, guerrilla urbanism, insurgent urbanism, everyday urbanism, pop-up urbanism, participatory urbanism.[19] These actions aim to politically expose the power dynamics behind control over public spaces as a spatial embodiment of democracy, though questions have been raised about what *counts* as a politically meaningful act of deviant spatial use (e.g., creating a temporary park in a parking lot as compared to a homeless encampment in the park).[20] In some cases these actions may start as apolitical disruption to the social system before coalescing into a political movement.[21]

Prefigurative urbanism shares some characteristics with other forms of social action, and most closely resembles prefigurative politics and DIY urbanism. All three realms of social action exhibit a distrust of the existing political and economic system, but in the case of prefigurative urbanism, adherents' actions reflect distrust not just of the government but corporations and even some philanthropic foundations. At first glance, some driving assumptions among these three realms of social action may appear to embody the core ethos of neoliberalism, particularly a belief in the ineffectiveness of government policies and bureaucracy, the superiority of market efficiency and innovation, and the moral rightness of individual responsibility. Closer examination reveals that they all reject neoliberal beliefs and values, but they do so in different ways. Prefigurative politics and DIY urbanism intentionally and explicitly work to undermine the institutions that perpetuate disparities in access to resources and opportunities, while prefigurative urbanism disregards the conventions that these same institutions implement and enforce. The fact that prefigurative urbanism is not explicitly political in aspiration or practice does not mean that the actions lack political significance. Their actions have political implications in that they implicitly or explicitly question legal and social norms, and their performative aspect underscores the actors' radical political inclination. At the same time, prefigurative urbanism actors do not attempt to dictate the meanings of their praxis to others, as they are constantly reconfiguring those meanings for themselves. In the post-Katrina New Orleans urban cultivation scene, growers embraced market opportunities and pursued social justice causes simultaneously without finding them mutually exclusive, partly because they did not narrowly define the scope of their praxis in advance. Nor did they make the *meaning* of their project their priority, even if they started with a set of aspirations and intentions for what they wanted to accomplish through it.

The individualized ethos of prefigurative urbanism also distinguishes it from the collective nature of prefigurative politics and DIY urbanism. Individual actors instigate change by creating a new system of production and consumption outside the existing market system. This is where prefigurative urbanism also departs from neoliberalism's operationalization of individual actions within a context of market choices and transactions.[22] Scholars and other commentators have critiqued neoliberal, post-disaster political economy for co-opting resilience, an approach that shifts policy focus to the capacity of individuals to endure and survive rather than work to minimize the risks of disruption and destruction.[23] While the practices of prefigurative urbanism risk inadvertently promoting the potential of individual actions over structural state intervention, the actions' involvement with others (e.g., neighbors, customers, other growers) distinguishes it from the consumer activism's impact that remains in the private realm. Prefigurative urbanism practitioners' innovations and adaptations also reject capitalistic "exchange values" or wealth concentration by enacting an alternative system of valorizing land, labor, and products. Thus, the New Orleans growers' actions go beyond the "quiet activism"[24] of seed savers; the growers use their practice to demonstrate publicly a possible alternative through spatial modification and cultivation of new social relationships and economic systems.

Prefigurative urbanism does not require the presence of a disaster or other form of social disruption to emerge as a form of social action, but this mode of action does resonate distinctly during times of crisis (Figure 6.1.). Disasters result from and intensify existing conditions of institutional failure and make them more salient; the urgent social conditions they produce also highlight the need for immediate change on a timeline typically impossible for social movements to achieve. Disasters disrupt the physical and social fabric of urban life, creating urgency and making acute the residents' need to figure out what to do next. In this context, Ann Swidler argues, our relationship to cultural practices, including mundane everyday action, gains saliency and coherence when we recognize that something needs to change for us to move forward or to prevent future disasters.[25] Prefigurative urbanism "makes more sense" to some individuals under certain social, economic, political, and cultural conditions, such as institutional failure, distrust in the system, and social norms that tolerate ambiguity. Actions that might quickly attract scrutiny and be forced to terminate under more strict governance and social surveillance regimes can be established and persist, allowing practitioners to see the possibility for change in the world.

Prefigurative urbanism thrives in a transitional social context, because shifting norms and institutional priorities create opportunities to experiment and adapt as actors continue to prefigure the alternative, especially when they operate in areas neglected by those in positions of economic and political power. Nevertheless, implementing something that does not easily fit within existing legal and social categories requires actors to creatively work around constraint, often with only limited support. Even after actors manage to establish a practice of prefigurative urbanism, they must continue to operate in shifting social environments. Some of these changes are independent of prefigurative urbanism. But if the practice is effective, it could gain the attention of the public and

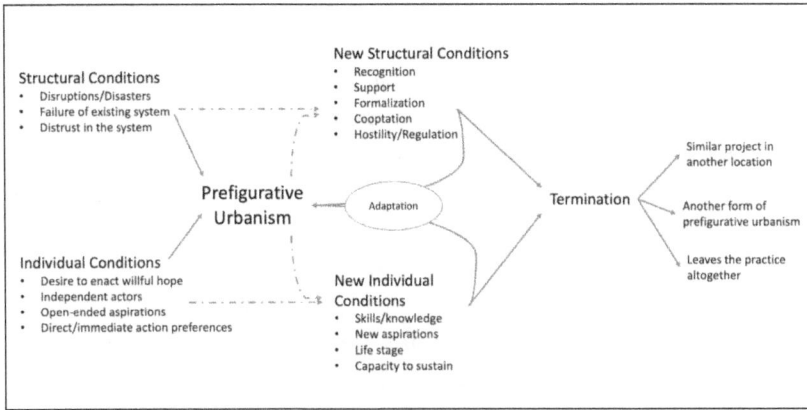

Figure 6.1. Emergence and Trajectory of Prefigurative Urbanism in Relation to External and Internal Conditions in Times of Disaster.

regulators, and others could begin to ascribe their own values to the practice. A practice might be coopted, ignored, supported, or challenged as a result. The actors must continue to adapt to these changes, but their decision-making remains focused on immediate priorities or outcomes. Over time, practitioners of prefigurative urbanism may sustain, terminate, or reorient their practices, but the ends-effacing and individualized nature of their practices makes it difficult to assess whether the action has been successful, or what would constitute "success." Prefiguring the alternative for any period at any scale may perhaps be considered a success, but how impactful are these actions in bringing about larger social changes beyond each project?

Stepping into the Void: Responding to Institutional Failures

There were plenty of reasons to be hopeless in post-disaster New Orleans, not just in the immediate aftermath of the destruction. The failures of the federal and local governments were not limited to the period immediately before, during, and after the August 2005 disaster; they extended into the months and years that followed.[26] It soon became clear that the *new* New Orleans would be shaped by neoliberal policies that emphasized free market-solutions and individual responsibility.[27] This version of "disaster capitalism"[28] invited newcomers, rather than long-term residents, to rebuild the city via gentrification and displacement, and it exacerbated existing residents' vulnerability to future disasters.[29] This neoliberal vision of hope for the city unapologetically embraced *new, outside* ideas, people, and money by loosening regulations and providing tax incentives. The policies encouraged private corporations to "experiment" with new ideas, from education to housing, while forcing individual residents to navigate a bureaucratic maze, in the name of fraud prevention, to prove themselves worthy of returning.[30] The insulting narratives of experimentation and innovation framed the city as if it were a "blank slate" waiting to be reinvented, undermining the capacity of long-term residents to rebuild their communities according to their own vision.[31]

Into the void created by the failure of the state and the market stepped nonprofit organizations and philanthropic foundations funding these neoliberal efforts.[32] Their presence in the city commodified the trauma of poor, Black New Orleanians while capitalizing on white saviors' desires.[33] Some high-profile scandals[34] involving local nonprofit organizations engaging in fraud or being infiltrated by an FBI informant later validated residents' skepticism toward the empty promises of the nonprofit industrial complex.[35] The proliferation of nonprofit organizations, especially in slow-recovering areas such as the Lower Ninth Ward, created its own sense of hopelessness by emphasizing the scale of the challenges and signaling the impossibility of addressing them. This was further complicated by the racial and class disconnect between low-income communities and nonprofit organizations, especially when the dominant faces of these organizations were white newcomers.[36]

Meanwhile, the grassroots activist movements that had been embedded in the community continued to demand justice, transparency, and reform by calling attention to the long-term historical roots of these problems. Their focus was not only on bringing displaced residents back or rebuilding what was lost, but also on advocating for the dismantling of the racist, capitalist, patriarchal systems that drove the disaster's disparate impacts.[37] The activists successfully pushed back against the neoliberal redevelopment agenda through coalition-building across race, age, and central grievances, from securing wage theft protection for workers and sweeping police reform to resisting the construction of a new landfill in the city.[38] They called out the neoliberal recovery agenda for trying to ignore or silence their voices and resisted the "community resilience" narratives that attempted to co-opt their movements.[39] Nevertheless, the city's public education, housing, and healthcare systems all became privatized, and the rapid pace of gentrification specifically threatened to permanently displace many long-term Black residents.[40]

In the context of structural conditions that could have induced collective hopelessness, urban cultivation emerged in post-Katrina New Orleans to activate hope at a small, practical, and manageable scale. Swidler argues that our cultural repertoire expands when institutions fail to guide our decision making, big or small, as we try to decipher what's right or wrong, or what is normal. In post-Katrina New Orleans, there were many institutional attempts to address "the problem," be it market-oriented redevelopment, charitable and philanthropic assistance, or social justice movements. Of these, social justice movements could have resonated with urban cultivation as a form of prefigurative politics, but those who initially began establishing cultivation projects did not situate their practice as part of a movement, and decidedly took an individualistic approach to enacting these changes. Cory reflected on how the post-Katrina urban cultivation scene did not forge connections to the city's existing social activism around racial justice:

> There has always been Black food justice organizing happening in New Orleans, you know, from gardeners to people [who] are in leadership positions in these organizations. Pam Broom, for instance. And there's a long tradition of Black gardeners and urban farmers in the city. Jeanette Bell is one who is still growing and has been growing most of her life, you know? I think that a piece of the phenomenon of the post-Katrina urban agriculture movement was that a lot of the people doing it were non-native New Orleanians and post-Katrina transplants. And I think that a lot of it wasn't necessarily happening in the context of multi-issue racial justice organizing. There's plenty of organizations [working with such approaches in addressing social justice]. . . . But I think that a lot of the urban farming work was a lot more rooted in supporting and building food systems, supporting local organic food systems and also creating that. And that is a movement that I think largely, at least in New Orleans post-Katrina, was not really connected in sort of organizing ways with those other movements.

Urban cultivation practices in post-disaster New Orleans did not develop direct connections to post-disaster grassroots movements for several reasons. First, the growers' aspirations were not grievance-driven, nor did they aim to create a collective identity or organization around their shared practices of growing food in the city. Second, given their commitment to a prefigurative approach, urban growers did not find it necessary to define their practices or predetermine the end results. Finally, communal or public cultivation practice had become less prominent in the city by 2005, except in the Vietnamese American communities in the East, making it less likely for the new growers to learn about historical precedent or the grassroots activists to connect their mobilization to urban cultivation.

The institutional failures of post-Katrina New Orleans created a unique set of opportunities for the growers' unconventional practices. The city's neglect of Black communities and working-class neighborhoods resulted in little oversight of activities taking place in these areas, exposing the residents to the risks of illegal dumping, low blight regulation enforcement, and poorly maintained infrastructure, such as non-functioning streetlights and unpaved potholes. But this same lack of attention also meant that someone starting a garden or a farm in these neighborhoods was less likely to receive a citation, even if their actions did not officially align with existing zoning codes. This would change during the second half of the decade, when urban cultivation began attracting the attention of neoliberal growth, just as the number of growers and the lots being cultivated expanded. The new attention brought with it new opportunities, but also new attempts to co-opt urban cultivation into the redevelopment agenda, or worse, to displace growers from the cultivation sites. By 2018, speculative landgrabs began to threaten some cultivation projects located in neighborhoods that were expected to "turn over," confirming growers' suspicions that urban cultivation was at odds, not in alliance, with neoliberal urbanism.

Urban Cultivation and Gentrification

The complex relationship between urban cultivation projects and the gentrification of New Orleans best illustrates the potential and limits of prefigurative urbanism in a transitional city. By the time the city observed the tenth anniversary of Hurricane Katrina gentrification had become a major point of concern in many parts of the city. Stories of the proliferation of Airbnb short-term rentals in the Bywater neighborhood, or a proposed condominium development in the Holy Cross section of the Lower Ninth Ward by the Mississippi River levees, were widely regarded as inevitable signs of gentrification in a city whose demographic composition had changed significantly over the decade. Repopulation rates differed drastically along racial lines, with the city's Black population declining by nearly ten percent and the white population growing by more than five percent.[41] These diverging numbers reflected stark disparities in residents' ability to access the resources necessary to evacuate or return to the city. But residential demographic transition, including an influx of white residents, was primarily concentrated in what Richard Campanella called the "white teapot" areas along the Mississippi riverbank and around Uptown, as well as in the Esplanade Avenue corridor to the northwest of the French Quarter.[42] Neighborhoods such as New Orleans East and most of the Upper and Lower Ninth Wards were showing little signs of actual development, even well into 2023.

The urban growers noted these changes, and they expressed varying levels of concern, based on their location and their access to the land they cultivated. In general, they had mixed feelings about the relationship between gentrification and their cultivation projects. Gentrification came up without specific probing in my conversations with the growers, during both the initial and the follow-up interviews. A common response among growers was that they might have contributed to gentrification, but they were also negatively impacted by it, not unlike the artists who moved into the lofts of the Lower East Side of Manhattan in the 1960s and 1970s as early gentrifiers and were themselves later displaced.[43] Joel, for example, reflected on the first garden he and his growing partner worked on around 2011. He wondered out loud if their starting a garden could have contributed to the gentrification of that particular block:

> Yes, we had definitely mixed feelings about that, about starting [the garden]. At first it was a great thing, bringing the neighborhood together. . . . But [we created] two really nice, manicured, good-looking gardens with

chickens and signs and two little hoop houses, [and] people that drove by started seeing it as a more appealing neighborhood. And then developers [began to show interests in the neighborhood]. The gardens would be largely responsible for the neighborhood's feeling, I guess, more safe and more appealing and more welcoming to newcomers.

Even though Joel felt that the garden could have invited new attention to a particular block within a particular neighborhood, he also saw it as a part of the urban developmental cycle happening across the city.

In short, many of the growers did not see urban cultivation as the *cause* of gentrification, but they allowed that it could have guided its expansion by prefiguring what appeared to the developers and newcomers as an *improved* space. More than one grower in this study terminated their garden when they saw that the people the cultivation space had been designed to serve were no longer there, as was the case for Joel's first site. During its operation, however, the garden provided food and a social space for the long-term residents nearby. In New Orleans, development-driven threats to urban cultivation did not spur the kind of collective mobilization to protect a space that have been observed around community gardens in other cities.[44] This could reflect the difference between community gardens and the type of urban cultivation studied in this book, because these gardens and farms were implemented as a form of prefigurative urbanism and were being managed independently. The growers did not improve formerly blighted spaces with an intention to increase property values, especially at a time when redevelopment was not in sight. It is also unlikely that any one garden or farm played a significant role in encouraging development of a neighborhood or even a block. Nevertheless, the growers felt conflicted about the correlation, if not causality, between their project and the neighborhood changes.

Changes came quickly and dramatically in some areas of the city. This was particularly the case for growers who started their cultivation projects during the recovery or transitional periods. In 2018, Shannon observed the pace and the scale of the changes in the city since she had begun working on urban cultivation in the city:

I've seen the city change so much in seven years. I've been here longer than seven years, but just in ag. Then specifically agriculture in the city changed a lot in seven years. When I first started, at least with the folks that I was talking with, it was under the radar [and people did not suspect] that agricultural land could be really intensely threatened, because, at least in my experience, there were so many abandoned lots. There were so many other things that the city was thinking about at the time. . . . At this point, where we are, I don't know. I don't think we know any gardeners who haven't lost land. I think it's pretty palpable now. It's pretty right there in the front that agricultural land loss is tacked onto larger development pressure and gentrification.

As Shannon looked to the future, she was uncertain about how to recognize and reorient the relationship between urban cultivation and gentrification, because she saw urban growers as simultaneously being "the gentrifiers and the gentrified." The community that one of her main projects had engaged with in Mid-City was in the process of being displaced, and she worried that this would fundamentally defeat the central mission of the project:

> We really don't know what our future is. You can't predict the outcome of something that you really don't know. You don't have many examples of it. But we do know that if we don't think people who are [partners] of the [project's] community can't stay here, then there's no point in doing the work.

The growers' relationships with their surrounding communities remained complex. Most growers had developed loose, mostly positive relationships with their neighbors, regardless of their primary aspiration. It was inarguable, though, that the very presence of urban cultivation projects on lots rendered vacant by Hurricane Katrina symbolized displacement. State and federal recovery policies made it exceptionally difficult for Black New Orleanians to return and rebuild.[45] In that context, the establishment of gardens in place of homes denoted a departure from the residential communities that once had been there. David explained why he does not use the term "urban farming" to describe his cultivation project:

> I don't use the word "farming" down here. People are definitely afraid of it. So I don't use "farming." I don't use "agriculture" anywhere in the

Lower Ninth Ward, even though I got one acre. We could call it a farm as a name, but I grow food, I don't farm food. I don't look at it as a farm. And it all goes back to the first redevelopment that the city put out, where it showed all the Lower Ninth Ward as one big green spot. People are very afraid that they would end up losing more properties than they were able to own back to farming, which is what this property started as before it was developed into housing.

The potential association between urban "farming" and the city's since-discarded "green dot" shrinkage plan could have sparked a hostile response from long-term community residents. Even apart from the "green dot" controversy, David suspected that the language of "farming" or "agriculture" conjured for Black New Orleans sentiments distinct from the bucolic, virtuous, and white-dominated "agrarian imaginary."[46] In the American South, specifically in Louisiana, the agricultural industry is deeply intertwined with the histories of slavery and of state-sanctioned racial violence, much of which continues today at the Angola prison.[47] Beyond being a semantic choice, then, the description of the practice as "urban agriculture" or "urban farming" in the public discourse, even if many of the growers themselves did not actively use those terms, likely resulted in many long-term residents in Black communities to distance themselves from these new urban cultivation projects.

This did not mean, however, that Black folks in the city categorically rejected the idea of urban cultivation, as evidenced by the historical precedent of communal local food provisioning described in Chapter One. Even many of those residents who would have rather had their old neighbors back or seen a home being built on the lot preferred a garden over blight, vacancy, or a developer "flipping" the property. Margee described her relationship to the community surrounding the cultivation site in the St. Roch neighborhood as "the best situation that I've ever had in an urban garden." She continued:

People will come and sit with me in the morning and drink their coffee, and we've had people bring health books, if they heard that we had some sort of ailment or whatever. . . . I think that if there was a lack of housing or something, and we were doing [the cultivation project], it would probably not be a very warmly accepted thing. But I think the fact that there are lots that are pretty dumped on and stuff [makes them feel differently]. People will ride their bikes by and just be, like, "Thank you!"

Yet she and her growing partner at the time recognized that their practice must consider their relationship to the residents, because they did not live in the community. When an idea for holding an event for a local organization at their site came up, they eventually turned it down because, "it's not our neighborhood, we don't live in that neighborhood. And so bringing in a bunch of people of our demographic out of nowhere could potentially feel weird." Such an approach exemplified how the growers managed their role in and relationship to a community that was experiencing gentrification at variable speeds.

Regardless of any ambivalence neighbors may have originally held toward an urban cultivation project, the premature or unexpected termination of a project did negatively impact a community that had already been failed repeatedly by federal and local governments, the private sector, and nonprofit organizations, all of whom promised changes and rarely delivered. In post-Katrina New Orleans, long-term residents continued to be subject to the collective "secondary trauma" of being failed by these institutions, some well-intended and others intentionally exploitative.[48] The termination of an urban cultivation project, whatever the reason, was likely to be interpreted as another piece of evidence of abandonment or disparity. Rob described the nearby kids' response to the decision to discontinue Our School at Blair Grocery as, "What the fuck, you guys leaving? That's it?" He pondered whether the decision to terminate the project after ten years ended up validating neighbors' initial skepticism:

> Because we lasted so long, people got over that to a certain extent, [by the time] we left. I wonder to what extent they [now] think, "Hey, well, it turns out we were right in the beginning because you do have the privilege just to get up and walk away but we're still stuck here."

Most of the cultivation projects studied in this book were located in Black communities whose generational trauma, caused by structural and social-racial injustices, were only exacerbated by the 2005 storms. This context meant that unfulfilled promises carried a particular sting.[49]

While I did not interview nearby residents of the terminated projects, the growers admitted that some of their neighbors had expressed a sense of abandonment and dismay at their departure, especially if the growers had been working in a space for more than a few years. Ironically, the gardens and farms that were least at risk of displacement from gentrification had fewer neighbors nearby, or the community itself was more transient and the growers did not forge a deep enough relationship with them to result in deep sentiments of betrayal.

The impact of gentrification on the growers themselves cut both ways. As illustrated in Chapters Four and Five, land insecurity intensified as speculative landgrabs took off in "up-and-coming" neighborhoods, but the influx of newcomers and the thriving new food economy expanded the opportunities of growers involved in commercial sales. It also introduced a larger volunteer labor force and a new cohort of potential growers. The increasing prominence of post-disaster transplants to the city during the redevelopment period created a sense of turnover in who was growing and why. Shannon saw the presence of transplants as a significant challenge for the future of the urban cultivation scene in New Orleans:

> There's a mix of people who are in agriculture, but it's a lot of transplants who are growing in the city, and it makes it a very precarious conversation. It's like, how do you talk about urban land loss or agriculture because there's privilege in all of that as well. The worrisome thing is, if we don't find out a way to have that conversation somehow, we're going to lose land. We're just not going to not have land. We're not going to have agricultural land pretty much at all pretty soon.

Erin, who worked with several cultivation projects between 2008 and 2012, responded to my observation that starting and working on a new cultivation project in 2008 seemed to have been a different experience from how things were in 2015:

> That's a hard thing for me to answer. 'Cause I kind of have had a really, intense personal conflict around that idea of "Am I an old-blood here or am I actually the first wave of gentrification?" And the answer is I'm probably the first wave. But had it not been for me it probably would

have happened anyway. And it was happening. I was there, and part of the flow [that started the scene].

Like Erin, most growers did not view urban cultivation as a practice with the capacity to either drive or prevent gentrification, at least not by itself. Reflecting their commitment to prefigurative urbanism, growers typically continued operating their projects until the threat could no longer be ignored. In some cases, bureaucratic uncertainty over land access sometimes took a long time to sort out, which bought the growers a few extra seasons before they ultimately lost their land. In other cases, growers were caught off guard when a land ownership dispute or lease negotiation resolved more quickly than anticipated, resulting in their being displaced from the space with little notice.

In keeping with their commitment to an individualized practice, the growers did not develop a cohesive response to the problem of development-driven land insecurity. Prefigurative urbanism's diversity and focus on immediate, tangible actions proved a constant challenge to those, like the Greater New Orleans Growers Association that tried to mobilize the growers, especially when it came to addressing the larger, structural changes taking place in the city, such as gentrification. Prefigurative urbanism is most effective at producing micro-level social and spatial transformations, but these types of changes cannot by themselves counter the forces of neoliberal redevelopment or racial capitalism.

It must be noted here, however, that prefigurative urbanism also resists co-optation by retaining its distrust of the existing political and economic system. Scholars and activists have called out the co-optation of terms such as "sustainability" and "greening" as buzzwords to justify the redevelopment of low-income communities and communities of color who have already suffered decades of disinvestment.[50] Thus, while they may not form a substantial or coordinated resistance against the political power elites and their agenda, the growers in New Orleans were not unaware of or complicit with efforts to co-opt, capitalize, or redefine their practice. They focused their energy on what they could do "now, concretely," rather than plan for an uncertain tomorrow. And they also took advantage of failures within the system such as bureaucratic bloat and incompetence, which allowed them to persist without being concerned about the legality or normative acceptability of their actions.

Because each case was so distinct, sometimes these strategies paid off, sometimes they didn't. And the same strategy may not work repeatedly as the institutions and the circumstances under which they operated kept changing.

The urban gardens and farms discussed in this book played complex roles in their communities that cannot be easily boiled down to a singular purpose, such as sustainability or food security. In fact, despite its common portrayal in the scholarship, urban agriculture is neither exclusively radical nor neoliberal; neoliberalism creates opportunities for urban agriculture, yet those same opportunities also hinder its potential.[51] The extent to which urban cultivation actively engages with either force, enabling gentrification or resisting its impact, for example, has not been empirically explored longitudinally. Findings from my research indicate that urban cultivation practices are both reactionary and prefigurative, and their impacts are most effective at a hyperlocalized level for a short period of time. Geographer Lindsay Campbell argues that urban agriculture's benefits extend beyond food production capacities and ecological benefits, as urban cultivation spaces act as "*assemblages* of the built environment, biotic actors, human labor, and institutions."[52] It was their liminal placement between these competing urban forces that enabled the growers to innovate, experiment, and adapt to demonstrate possible alternatives. In the remainder of the chapter, I further explore the role that prefigurative urbanism plays in transitional cities with a focus on individual dispositions and historical and cultural contexts.

Reflection on Life Stage and Privilege in Prefigurative Urbanism

Beyond their personal dispositions, many of the individuals who started new public cultivation projects in New Orleans were at a transitional stage in their lives.[53] The two most common age groups among the growers were those in their twenties and in retirement; the former in search of a career path and the latter seeking their next journey. Post-Katrina transplants were literally in transition, spatially and socially, having arrived to the city while in search of a new opportunity for making a difference or finding a job.[54] These transitional moments elevate one's need to reorient their cultural routines and habits as a way to make sense of new or emerging circumstances, which includes taking on actions that help guide their transition.[55]

The dominance of post-Katrina transplants in the urban cultivation scene by the end of the decade requires us to consider factors beyond the actors' personal backgrounds and life stages: Who actually *could*, not just *would*, partake in prefigurative urbanism? Early in the recovery period, activists and academics alike chastised the "young, urban recovery professionals" (YURPs) who descended on the city with their "white savior complex" and naïve outlook on their potential impact. These young newcomers continued to arrive to fill the ranks of the city's thriving nonprofit industry. Critics argued that this type of philanthropic work allowed its practitioners to see themselves as morally superior saviors without reflecting on how these well-intended actions reinforced existing forms of privilege and power.[56] In reality, many of these transplants acknowledged the complicated nature of their role, yet they mostly expressed their ambivalence in terms of the city's changing culture.[57] Later in the decade, a different set of newcomers began to arrive, seeking opportunities associated with the city's revived hospitality and entertainment industries. These later transplants resembled what sociologist Richard Florida described as the "creative class" of young professionals who were drawn to urban environments that offered talent, technology, and tolerance.[58] Demographic shifts in the urban cultivation scene mirror these larger social changes in the city, though given their very small number, the growers remained an anomaly even within their own demographics.

To be sure, some post-Katrina transplant growers explicitly or implicitly used the "savior" framing in describing their urban cultivation aspirations. The sense that there was "nothing here" before Katrina in terms of urban cultivation was shared not just by the newcomers but also by long-term residents in all parts of the city, except for those few with historical knowledge of the practice. Some of the growers exhibited the "savior" stance when stating their initial aspirations, and a handful continued to describe New Orleans as being "behind" or "resistant to change." Again, such sentiments were not expressed exclusively by the newcomers. There were long-term or multi-generational New Orleanians, both Black and white, who, during my initial interviews with them, would lament that the city and its residents were too slow to embrace urban cultivation. Over time, however, most of the transplant growers abandoned notions of New Orleans as being "behind" as they gained knowledge of the traditions and legacy of local food provisioning in the city through their interaction with older long-term residents as neighbors, customers, or collaborators. Some growers also gained more critical perspectives through their exposure to food justice and environmental justice activism and scholarship, as I did in my own journey through this research.

Many of the growers interviewed for this book, especially younger transplants who started their cultivation projects with the *community rebuilding* aspiration, frankly addressed the problems of the savior stance. They candidly reflected on their positionality, focusing primarily on their race, class, age, and geographic background, and they acknowledged their learning curve of the social implications of their practice, sharing critical assessments of how they learned from mistakes or gained a more nuanced understanding of local history and race relations. Cory, who co-led the Foodshed Roundtable dinner conversations described in Chapter Four, reflected on where the conversation hit a roadblock:

> I think I started it, because I was critical of the ways that I was seeing, sort of the savior complex and gentrification intersecting with the local food movement in New Orleans. So I was trying to have some conversations around that, you know. It sort of grew on its own into this like a short-lived collective. I think that how I ended up stopping [was], we had some kind of tough moments around being a group of majority white people, and also pretty young majority white people. It was around the time of Occupy [Wall Street movement], too, I think. Majority white people with pretty decent racial politics, but also really young and sort of new to having conversations. And I think that when we bumped against some criticisms, we kind of had a hard time moving through them, you know, if I remember.

Scholars who write about Occupy Wall Street, in which young protestors camped out in public spaces protesting against wealth disparity in the US, often describe the movement as a prototypical moment of prefigurative politics in action. When Cory mentioned Occupy, however, they had in mind the movement's failure to effectively grapple with the internal racial and class privileges of the participants.[59] The challenges of conducting these conversations with people who struggled to sit with discomfort or were not experienced in facilitating such conversations, illuminates insularity and the underlying privilege of these social spaces, despite participants' good intentions to address broader structural injustices.

Growers' attempts to incorporate self-awareness of their racial and class privilege into their practices proved complex and subject to constant re-evaluation. Zach, a white transplant who bought a home in the Ninth Ward in 2013 and first began growing in his backyard before expanding to a couple of other lots in the area, had to revise his own visions for urban cultivation in light of his neighbors' vision for their community. As described earlier in Chapter Four, he tried to balance his permaculture practice with his neighbors' needs and values. Around 2018 he erected a fence around a new cultivation site that he had finally purchased from NORA, which was located across from a corner store. His initial plan had been to leave it open for the community to access, but he installed the fence after one of the older, long-term Black neighbors expressed concern that an accessible garden would attract unwanted loitering. Most other residents were fine with the fence, he recalled, but then a Black resident accused him of being racist for being exclusionary. He recalled thinking to himself, with a tone of humor, "I was told by Black people (in the neighborhood) to close it up . . . I wanted to keep it open!" Zach reflected on his relationship to the community and the broader political spectrum:

I mean, I'm coming in as an outsider, not knowing the relationships that are already intertwined. So we have an event here and two sisters get into a fight. That event was called Breaking Bread. I think it was after Trump was elected, and I just wanted to have something that was like a potluck for the community. . . . I think it's a realization that we're all just trying our best. This side or that side, there's just as much vitriol and anger and the side that I agree with, if you will, I guess, more 'liberal' for lack of a better word. But nowadays, liberals are just as evil and mean and menacing to the other side as they espouse to want to take down, you know?

As a relative newcomer to the community, Zach had to grapple with the complexity of the community's social relationships. Not everyone in the community agreed on everything, and the residents had their own history. He situated himself as being a part of the collective efforts to "just try our best," with an awareness that, as an outsider, his actions were subject to additional scrutiny and might never fully satisfy everyone in the community. Zach's concerns about the increasingly divisive political landscape, in the aftermath of the election of Donald Trump as US President in 2016, was also typical of the growers. Many were socially progressive and some even held somewhat radical political views, but their ideas about urban cultivation were grounded more in pragmatism than idealism. Those few who employed explicitly political concepts such as food justice or environmental justice exhibited varying degrees of commitment to the concepts in their actual work.[60]

Aside from the obvious issues of financial security discussed in prior chapters, growers' willingness to engage in practices that violated existing social and legal norms reflected disregard for the consequences that stemmed from a combination of personality and social positionality. To the extent that they considered themselves members of the community where they operated, they justified actions that defied norms as a way of meeting community needs that have not been, and will not be, met by the systems in place. They also accurately speculated, based on the long-term neglect of these communities, that the risk of regulatory prosecution was low, at least until the regulatory pressure intensified during the redevelopment period. Those who were newcomers were more likely to exhibit blasé attitudes toward the potential legal or social consequences of breaking norms, taking into account their racial and class statuses, New Orleans's infamously inefficient bureaucracy, and scarce public attention to urban cultivation. In other words, growers did not act necessarily as if they were "entitled" to do as they wished, but they assumed there would be few consequences for doing what they wanted. Engaging in prefigurative urbanism requires the actors to not be deterred by the prospects of breaking social and legal norms. In the case of the growers in New Orleans, such conviction resulted from a combination of their social status and personal temperament. To pursue being an urban grower, one must be comfortable doing something that most others would not, in addition to being able to afford to do so in the first place, both financially and socially. Yet by positioning urban cultivation as an alternative, new approach, rather than a continuation of the city's history of food provisioning, most of these new urban growers defined their practice through a position of relative privilege, rather than manifesting their vision as resistant to the status quo in solidarity with marginalized communities.

History Matters

Given the rising popularity of alternative food movements and other sustainable practices, the new wave of urban cultivation would have eventually arrived in New Orleans, even without Katrina. New Orleans has always incorporated newcomers and new ideas into its cultural traditions, if at a much slower pace than other cities. Without a major disaster, perhaps some of the long-term residents' skillsets and interests would have been supported through newly available resources. Local activists might have highlighted the historical precedents of urban cultivation practices in Black communities and reclaimed that narrative. Or perhaps "urban agriculture" would have been introduced as a new idea by a local NGO or government agency, similarly to how Parkway Partners in the 1980s presented community gardening to the city. Most likely, all of these scenarios would have happened simultaneously. Regardless, New Orleans's urban cultivation scene would have developed more organically, adapting to the economic and social landscapes of an extant, functioning city. This has been the case in post-industrial American cities like Detroit, Baltimore, and Philadelphia, where urban cultivation retained its presence in working-class neighborhoods and gradually expanded in association with grassroots movements as well as local political agendas.[61]

But this is not, of course, what happened with New Orleans. The slow and steady decline of communal urban cultivation practice in the city took place decades before the 2005 storms. Erasure of the legacy of urban cultivation in the city's Black communities resulted from a confluence of historical factors.[62] As described in Chapter One, the practice of growing food for themselves and others among the city's Black communities was in decline by the turn of the twenty-first century. Beyond those directly involved in gardening and farming in the city, New Orleans, including the local government, mostly remained ignorant of or uninterested in prefigurative practices in Black or immigrant communities. Mass displacement in the wake of Hurricanes Katrina and Rita exacerbated the loss of such memory with the older generation's departure, so many of whom did not return to the city to resume gardening or to share their memories and knowledge. Thus, when the new generation of urban growers began clearing lots to grow food, herbs, and animals in the city, they were not seeing indications of the earlier practices. The emergence and trajectory of urban cultivation as a form of prefigurative urbanism in post-Katrina New Orleans resulted as much from the social and economic transformation of the city after the 2005 disaster as from what sociologist Paul Draus calls the "slow-motion disasters"[63] of economic decline, population loss, racial segregation, and divestment from public infrastructure decades prior.

The distinct trajectory of urban cultivation in the city's Vietnamese immigrant enclave illustrates this point. The residents there were still growing food at scale and engaging in communal distribution into the 2000s. The insularity of this enclave and the immigrants' relatively recent arrival to the city necessitated that the community exercise economic self-sufficiency, with the Catholic church playing a central role in advocating for resources, including the land on which a sizable urban farming operation took place.[64] But these practices remained largely unknown to the rest of the city until Hurricane Katrina, and the practice might have gradually dissipated altogether because it was mostly elderly, first-generation immigrants who were tending the gardens. In the late 2000s, however, a series of challenges—including the floods, a proposed landfill, and effects of the BP Oil Spill on the Vietnamese shrimping community[65]—mobilized younger, second-generation immigrant activists to take up growing. They turned to urban cultivation practices as a milieu for community rebuilding, ultimately establishing the VEGGI Cooperative. Even in this community, therefore, the arrival of younger activists in response to Katrina in 2005 and the BP Oil Spill in 2010 played a significant role in expanding a practice that most of the younger people in the community were not otherwise set to inherit from the elders.

Enacting Hope through Prefigurative Urbanism

As should be clear by now, urban cultivation practices in post-Katrina New Orleans primarily took the form of prefigurative urbanism. It remains for us to ask: What role, if any, does prefigurative urbanism play in the city? Here, is it useful to compare the potential impacts of prefigurative urbanism with its most closely related mode of social action, prefigurative politics. In their review of scholarship on prefigurative politics, geographers Craig Jeffrey and Jane Dyson identify several critiques of this form of oppositional politics, particularly its tendency to occur in a selective, privileged social space for those who can afford to adhere to ideological purity.[66] Another line of their critique problematizes the failure of prefigurative politics to hold the state and capitalistic forces accountable for their role in creating the structural causes of injustice; without that vigilance, prefigurative politics remains vulnerable to co-optation. But Jeffery and Dyson also identify four specific ways through which prefigurative politics brings about change: scaling up into a larger movement; developing skills, knowledge, or resources; changing public attitudes; and fostering "affective importance," or a sense of collective efficacy.[67] Aforementioned Occupy Wall Street movement, for example, did not dismantle capitalism or fundamentally change the US's economic structure, but the movement did leave a mark in public consciousness through its "We are the ninety-nine percent" slogan. Long after the protestors were removed from Zuccotti Park, the public kept in their awareness a slogan that pointed to the severity of social disparity in American society. Subsequent movement organizing took place in the context of this new public awareness, which could have also indirectly contributed to the rise of the Tea Party movement on the right that galvanized around distrust of the government. Yet the Occupy Wall Street example demonstrates that long-term ripple effects of prefigurative politics may be disruptive in ways that diverge from, or are even in opposition to, the activists' original intentions.

The particular characteristics of prefigurative urbanism present similar, if distinct, possibilities and limitations for enacting social change. Actors who are engaged in prefigurative urbanism focus on immediate, concrete outcomes, making the approach attractive to those who want to see changes take place in convincing and meaningful ways, however small in scale. But unlike in cases of prefigurative politics or DIY urbanism, these alternative visions are not necessarily politically radical or intended to disrupt. Their actions *may* end up being figuratively and physically disruptive, with political implications. Nevertheless, to the extent that this happens, it is a side effect, rather than the primary motive. Practitioners' orientation toward pragmatism, rather than ideology, results from their paramount commitment to immediate direct action, which reflects the ends-effacing nature of prefigurative urbanism. As with prefigurative politics, this tendency directs prefigurative urbanism away from attempting to change or explicitly challenge the structural causes of the issues practitioners attempt to address. At the same time, prefigurative urbanism actors' pragmatism allows them to establish their practice quickly and without compromise, at least for the short term. Over time, the projects that make sense for a particular time and place will find ways to sustain themselves, in a process not dissimilar from the ecological process of evolutionary adaptation. In other cases, if the alternative future they prefigured has become widely accepted, the practice is no longer prefigurative; it now operates in the context of legal and social legitimacy, which poses a new set of constraints on the practice.

Prefigurative urbanism actors' preference for individual action retains and prioritizes a heterogeneity of practices over the formation of collective identities. This individualized approach offers practitioners flexibility in their ability to adapt or experiment in response to changing circumstances, but it also creates vulnerability to burnout, termination, or co-optation. These actors' adaptive process keeps their practice grounded in hyperlocalized contexts rather than oriented toward broader, more generalized outcomes, while allowing space for the actors to recalibrate their practice as they go. As such, prefigurative urbanism is more effective in offering what social movement scholars call *prognostic framing*, or suggesting concrete solutions and actions (e.g., to grow food in the city), without presenting singular or specific *diagnostic framing*, or explaining the issue and identifying its causes, on how and why urban cultivation may be the solution.[68] Prioritizing "what" over "why" may be especially pertinent in times of disruption and uncertainty, when the situation is quickly unraveling with little information on what is to come.

Prefigurative urbanism operates somewhere between the two opposing effects of optimism and pessimism by opting to exercise hopeful actions grounded in pragmatism. Sociologist Tressie McMillan Cottom defines "pragmatic hope" as the idea that "you will not always see the good ends of your good deeds, but you do it anyway."[69] With such hope, Cottom argues, we find ways to implement executable actions toward larger changes, "knowing from the outset that [we] are going to fall short of them."[70] In prefigurative urbanism practice, then, individuals activate a pragmatic and willful hope by manifesting a concrete vision of an alternative future. Unlike prefigurative politics or social movements more generally, prefigurative urbanism is not grievance-driven. Recognizing social problems and learning about their deep, complex root causes can motivate one to join a movement with the hope of making change, but it can also lead to despair when the problems seem too big or complex. Prefigurative urbanism enacts hopefulness by proving that *something can be done* in observable and immediately impactful ways. But this stance is not grounded in optimism. The growers discussed in this book were not optimists seeking utopia, and some of them critiqued other urban growers for being "naïve," "cute," or "idealistic," underscoring their disdain of blind optimism. In other words, prefigurative urbanism instills hope specifically in response to a sense of despair, especially the skepticism about the larger system's ability to change in an immediate future, by manifesting the kind of concrete changes that can be achieved in the short-term.[71]

The type of hope that prefigurative urbanism activates through direct, public action is distinct from the resilience-oriented solutions that neoliberal policymakers often tout in disaster-prevention policies.[72] Instead of either relying on "resiliency" or proposing policy changes that could prevent future disasters, prefigurative urbanism focuses on individuals' ability to directly work toward broader social change. In recent years, grassroots activists' efforts to represent and advocate for their own communities has become increasingly at risk of co-optation under the guise of "civic engagement";[73] well-intended social justice organizations have come under the scrutiny of activists and scholars for replicating racial and class hierarchies within their own movements.[74] Prefigurative urbanism actors' engagement with the public is not completely free from these pitfalls, as I have demonstrated in Chapters Four and Five, but the actors' balancing of pragmatism and willful hope results in a relationship with the public that is more practical and independent than idealistic or collective.

Over the long term, the individualized practices of prefigurative urbanism do create a larger, cumulative "ripple effect" that departs its practitioners' control or original aspirations. Its existence convinces some, if not all, of the public of the possibility of alternative uses and values for urban land. It encourages other individuals to take up their own prefigurative urbanism practices, related or unrelated to the original forms of action. It could develop into a social movement or become a kind of prefigurative politics with a distinct mission and political intention, though such transitions are not guaranteed, nor, as we've seen, were they pursued by most growers in New Orleans.

The establishment and persistence of urban cultivation as a form of prefigurative urbanism created new social and economic systems as the growers opted to operate outside of existing systems, continuing instead to experiment and innovate. Shannon articulated the challenges and potential of addressing complex social issues through the individualized actions of urban cultivation:

> I think it's hard, too, because we can't address all the things. One organization can't address all the things. I know I can't. I would just be exhausted. My head would be spinning if I know that all these social problems exist. I know that we can't just engage with gardening and then that's the end of the conversation, like "We did it! That was it! That was our thing that we contributed." At the same time, we're just human. How far do we have to reach? We're able to do our part. We're able to do what we're good at. We're able to do what our calling is to try to come up with a solution and there's this whole team of other folks who are doing their portion of it, that we can just work together in a network and try to really transform.

These long-term outcomes result from the interactive and collective impacts of individualized prefigurative actions, even without the actors' collective organization, but the actors involved do not dictate or predetermine the range and trajectory of these impacts. This is why prefigurative urbanism develops and persists at different paces and scales in different cities, depending on the pace at which economic, political, and social conditions are changing. While prefigurative urbanism's success can therefore be measured in terms of its immediate capacity to convincingly prefigure an alternative future, its long-term impacts require in-depth, longitudinal assessments.

Conclusion

Urban cultivation practices can, and often do, emerge and become sustainable as a part of political activism, but this was not the case in New Orleans.[75] Instead, the heterogeneous aspirations, evolving practices, and divergent trajectories of urban cultivation projects in New Orleans since 2005 provide us with a glimpse into a category of social responses to crisis that do not fit squarely into the prevailing, and competing, narratives about either "disaster capitalism" or "grassroots resistance." The growers' aspirations, practice, and trajectories were undeniably shaped by the city's emerging neoliberal political economy, but individual growers, many of whom were themselves going through moments of personal transition, forged their own paths to enacting pragmatic and willful hope rather than joining social movements or other forms of existing social actions.

What, if anything, did this approach to urban cultivation as prefigurative urbanism accomplish in the post-disaster city? The cultivation projects discussed in this book produced some concrete outcomes in the short term: they transformed physical space, produced and distributed locally grown food, explored new food distribution systems, and demonstrated the possibility of urban growing as a career or a way of life. While some of these consequences were intended, others were unanticipated byproducts of cultivation. These direct actions themselves held some meaning for both the growers and their neighbors, but they did not fundamentally challenge, let alone dismantle, the capitalist forces that were remaking the city or the global food chain, as was made evident by the growers' own vulnerability to gentrification. The long-term, aggregate impacts of prefigurative urbanism are less tangible and direct. Some cultivation projects terminated, voluntarily or involuntarily, while others expanded and spurred spinoff projects or even attempts to organize collectively. The evaluation of this form of civic action, I argue, should focus on how these direct actions *cultivated*, or prepared for growing, hope for positive change—change that the actors themselves could not yet fully anticipate.

What happened in post-Katrina New Orleans underscores the complex and variable role that prefigurative urbanism plays in the city. The actions of well-intentioned individuals could inadvertently facilitate or complement gentrification of the very neighborhoods they wished to help rebuild; new terms and faces of the practice could overwrite the legacy of pre-existing but under-recognized practices of the long-term residents in marginalized communities who used to prefigure alternatives in the past. In this sense, the previous iteration of urban cultivation in New Orleans took place in the context of "slow-motion disasters" of economic decline and perpetual racial injustice, in contrast to the spectacular and abrupt devastations caused by Hurricanes Katrina and Rita. Historical practices by Black and immigrant growers were overlooked and underestimated by the city's power elites, even as these gardeners prefigured food sovereignty and productive appropriation of undervalued spaces but remained contained within a particular generation and locale. By contrast, urban cultivation practice since 2005 occurred in the context of urban redevelopment, rather than decline, with transplants dominating the practice, informed by larger discourses surrounding sustainability and alternative food systems, rather than local history. Prefigurative urbanism reflects and interacts with the social, political, and economic conditions of a particular place and time, even if the actors are not explicitly politically engaged.

Conclusion

A lot has changed in New Orleans and the rest of the US since I formally concluded data collection for this research in 2018. Gentrification has intensified, New Orleans experienced an early and severe outbreak of COVID-19 in 2020, and Hurricane Ida brought a stark reminder of climate change on the sixteenth anniversary of Katrina in 2021. As the US braced for another coronavirus surge, a polar vortex froze everything in the city for nearly a week over the 2022 Christmas holidays. Not surprisingly, urban growers in New Orleans continued to adapt to these and other uncertainties, building on their experiences from the previous decade to find ways to continue to *do something, now.*

During the initial year of the global pandemic in 2020, Margee, the executive director of SPROUT at the time, distributed food to community members who were food-insecure and organized plant distribution across the city with other volunteers so that people could start growing food themselves. Jamal adjusted his growing practices at his project, Supporting Urban Agriculture, in response to the unpredictability of restaurant sales, and took on a teaching position at a charter high school in New Orleans East for a semester. But despite the damages caused by Hurricane Ida, which shut down his operations for at least six weeks, and the polar vortex that froze his winter crop in 2022, he kept on growing. In 2020, Joel's business model for Paradigm Gardens took a major blow from the event cancellations and restaurant closures caused by the pandemic, but Joel knew his project would ultimately survive because it had no paid employees and minimal maintenance costs beyond his own labor. He and his business partner, A'Keem, meanwhile, decided to try something new: an alternative school for the families whose children were stuck at home for virtual learning during 2020 and 2021. These growers continued to let their work speak for itself as they experimented and adapted to prefigure an alternative.

The urban cultivation scene in New Orleans continues to change in terms of who is growing, and where, how, and why they grow. While the new growers who entered the scene after 2015 are not the subjects of this study, some of their projects sprouted in the very same spaces vacated by earlier projects, others at entirely new locations. As before, many of the new projects are either being started or taken over by transplants, but some of these new projects exhibit more explicit political aims than their predecessors. Solitary Gardens in the Lower Ninth Ward, for example, is a prison abolition movement demonstration project by jackie sumell that works with incarcerated people in solitary confinement to design garden beds the same size as their cells, as an artistic prefiguration of "a landscape without prisons."[1] Speak Easy Farm, located in the St. Claud neighborhood in the Upper Ninth Ward, has been working to create an inclusive space through workshops and community events focused on herbalism, ecological sustainability, and the local cultural and histori- cal significance of growing plants and food in the city. The long-term outcomes of these new projects remain to be seen, but their emergence and trajectories will continue to reflect and interact with the changing economic, social, and political conditions of the city.

Approximately one-third of the urban cultivation projects or grow- ers who participated in the original interviews continue to operate in the city as of early 2024. Since 2015, Pamela has worked with NORA, Parkway Partners, and a network of advocates to improve how grow- ers successfully access and benefit from NORA's Growing Green lease program. She later began working for NewCorp, Inc., a local commu- nity development financial institution (CDFI), where she developed the FARMacia project in the Seventh Ward. Designed as a wellness hub for the community. The site's gardens and open-air educational spaces are coincidentally located on the former property of Dr. Thelma Coffey- Boutte, the first Black woman physician to practice in New Orleans (from 1937 to 1985). Jeanette also continues to grow in multiple spaces in hopes of encouraging and educating the new generation of growers. She had also been welcoming volunteers from partnering universities and organizations that kept returning year after year, and her work has been recognized in books and media reports on urban cultivation. In 2024 she lost access to most of the lots she had been leasing through the HUG program, but she has since pivoted to focusing on tending the orchard and propagating the plants on the remaining spaces. The work of these growers continues to reflect their *urban cultivation expansion* aspiration and focuses on demonstrating how to grow in the city to those who might not otherwise be interested.

The growers with the *community rebuilding* aspiration remained engaged in prefigurative urbanism, as they reflected on their own relationship with urban cultivation and the community. David has been adjusting Capstone's operations to focus on aquaculture and the orchard to reduce the physical demands of maintenance, both in response to his own health issues and the loss of volunteers during the COVID-19 pandemic. During his decade of growing in the Lower Ninth Ward, he observed several hastily built projects that did not last. In contrast, David carried on his work with no major funding but with the support of the local community. Some of the properties where Capstone planted citrus trees now have homes on them; but many of the new owners have kept these trees. On the other side of the industrial canal, Zach could point to at least a dozen orchards he had planted and stewarded around the neighborhood. He did not seem concerned about the fallen branches and leaves all over his own cultivation space—damage from 2022's polar vortex—but he was contemplating where his journey with urban cultivation would take him next. With a tone of admiration, Zach observed that the new generation of growers in the city had more clearly defined visions for their projects compared to him and his fellow growers when he started. He did not have immediate plans to stop doing what he was doing, but the long-term plans remained open-ended.

Some of the growers' organizing efforts have persisted since the formation of the Greater New Orleans Growers Association (GNOGA) in 2012, and Margee and others from the association continued to pursue policy work on their own beyond the association. Margee ran for the position of Louisiana Commissioner of Agricultural Forestry in 2018. Her pursuit of political office was a rare departure from the typical form of prefigurative urbanism practice among the growers in New Orleans, but her ambition was in many ways an extension of her work "on the ground." Her campaign attempted to prefigure changes in the priorities of the influential agency that oversees the state agricultural and timber industries to better reflect the needs of small-scale growers in cities and rural areas. Even though Margee ultimately lost the race, garnering twenty percent of the vote against an incumbent who had held the position for fourteen years, her campaign brought media coverage to a position that typically does not receive much public attention. She has since resumed her role as a grower and an advocate for urban cultivation in New Orleans, building on the relationships with the incumbent and other candidates she developed during her campaign. Shannon worked with the National Young Farmers' Coalition to advocate successfully for urban agriculture to be included in the 2018 Farm Bill.[2] Despite all the policy work and organizing, however, her central aspiration remains to be a grower. Having not yet succeeded at securing land for the long term in New Orleans, despite several attempts, she described herself as "a farmer without land."

This book has examined why and how urban cultivation emerged in the specific context of post-Katrina New Orleans, as well as how the cultivation projects evolved over a decade or so following the storms. While certain aspects of these growers' stories are unique to the circumstances of their place and time, the growers' experiences of trying to start and sustain their projects in a transitional city have implications beyond New Orleans. Urban cultivation remains popular across the US at a moment when many cities are undergoing economic, political, social, and ecological transformation.

What Next for Urban Cultivation?

"Urban agriculture" is no longer a radical concept in the US, as evidenced by the establishment of the US Department of Agriculture's Office of Urban Agriculture and Innovative Production in 2018. Urban agriculture has been increasingly incorporated into urban planning and municipal policies, albeit without a clear definition of what constitutes "urban agriculture." Community gardens, school gardens, urban farms, and backyard gardens are all distinct forms of growing plants and animals in the city, with varying intentions, forms, and scales. More generally, the perception of urban agriculture as a panacea for a range of social issues, from health to environmental sustainability, education, and under-employment, persists. A capitalist valorization of localism and marketing concepts like "farm-to-table" perpetuate the idea that urban agriculture is inherently communal and equitable, despite scholars and activists calling these assumptions into question.[3] Policies that promote urban agriculture without acknowledging what it takes to engage in cultivation in the city will either be ineffective or selectively supportive of the type of projects that fit the green gentrification framework.[4]

The mere act of adopting policies that seem supportive of urban agriculture does not produce more gardens or farms in a city, because the policies cannot be implemented without cross-agency coordination and access to resources.[5] Urban growers' diverse practices fall under a range of public agencies' purviews, from water, zoning, and permitting to health and education. Without intentional coordination across these agencies, growers will continue to have to navigate the multitudes of local agencies that regulate different aspects of their practice. This is additionally complicated because urban agriculture policies rarely reflect the heterogeneity of cultivation practices, and the agencies tasked with regulating the practices often do not understand them or have a narrow vision that excludes many forms of urban cultivation. The result is that regulatory oversight can become unnecessarily restrictive or exclusionary toward certain types of urban agriculture, which may encourage growers to operate in legal gray areas rather than seeking legitimacy. This is especially the case for growers engaging with urban cultivation as a form of prefigurative urbanism. Without strong advocates on the ground to demand and facilitate policy implementation, therefore, urban agriculture policies do not activate or sustain cultivation practices—at least not in all forms. Such policies are especially unlikely to work in gentrifying cities, where competing priorities such as commercial and residential development take precedent over urban gardens and farms. After all, most policymakers do not consider urban agriculture essential to the function of a city, despite all of the hopes ascribed to the practice by the public and, increasingly, urban agriculture policies. Furthermore, policies that do not proactively address the historical racial injustices that underlie the changing demographics and intentions of urban growers would likely exacerbate existing social inequalities, as evidenced by the legalization of cannabis in some states, leading to an intensified disparity in economic and criminal justice outcomes along racial and class lines.[6]

The two models of urban agriculture that are establishing themselves most quickly embody the competing forces of urbanism: neoliberal capitalism and grassroots activism. On the one hand, controlled-environment agriculture (CEA) and other technology-mediated forms of urban food production have captured the imagination of investors, especially those with the savior mindsets. These entrepreneurial growers tout the potential of computerized water-based growing systems, typically set up in formerly abandoned warehouses or on rooftops, for high-density, year-round food production.[7] Whether they are more ecologically sustainable than conventional in-ground production remains inconclusive, given their energy- and resource-intensive nature.[8] The supporters of this model of urban agriculture align it with a market-driven, neoliberal imagining of an alternative future while conveniently overlooking its practical and ethical implications. Working in this type of agricultural manufacturing requires high-tech skills; the infrastructure requires substantial capital for the initial development; and types of food grown, such as microgreens and lettuces, may not match the cultural and social needs of communities experiencing food insecurity.

On the other hand, urban agricultural projects that are centered around Black and indigenous identities and histories have been expanding their work to advance environmental and food justice causes across the United States. Grower-activists and groups such as Karen Washington and Soul Fire Farm in New York state, the Detroit Black Community Food Security Network, the Black Dirt Farm Collective in the Mid-Atlantic region, the Urban Growers Collective in Chicago, and the Southeastern African American Farmers Organic Network are just a few examples of cultivation projects and organizations addressing the intersection of land, food, and environmental injustice in their practice. The collaborative networks between these projects are geographically expansive, connecting urban and rural small-scale growers across the country and globe as a part of the "new agrarianism" movement.[9]

Given these trends, media and academic research on urban agriculture continue to focus on practices that align politically with either of the opposing forces, neoliberal remaking of the city and grassroots community activism, while overlooking forms of urban cultivation that do not fit squarely into these explicitly capitalistic or political models. While food justice scholarship has successfully shed light on the activist-growers who use cultivation projects as a form of prefigurative politics or a part of more conventional social-movement organizing, sole emphasis on this type of growing risks casting all urban cultivation practice as being inherently political, while disregarding the transformational potential of actions that are not explicitly political. Moreover, most public attention to urban cultivation projects focuses on projects that are being established or have managed to sustain themselves for several years, without recognizing how many gardens and farms never fully launch or go fallow even after decades of operation.[10] Furthermore, the growers engaging in prefigurative urbanism cultivation are either not being recognized or categorized into one of the two existing opposing categories in scholarship or media reporting, without critical assessment of how they see themselves and their projects. This book represents an attempt to correct these narratives by highlighting the breadth of urban growers' practices, motivations, and trajectories.

The most pervasive misconception about urban cultivation concerns the resources required to establish and sustain a garden or a farm. The celebratory expectation that urban agriculture will somehow address food insecurity rarely considers the cost of land and labor that keeps urban cultivation beyond the economic reach not just of low-income residents and people of color, but even of white, middle-class newcomers, as this book demonstrates. In practice, growers and urban cultivation organizations make ends meet through a combination of volunteer labor, side hustles, philanthropic grants or donations, local governmental subsidies, and by identifying markets willing and able to pay prices high enough to subsidize feeding their low-income neighbors. Curiously, the public ignorance of the hidden "cost" of urban agriculture is at odds with its narrow valorization as a progressive "food production" practice. In many cases, the growers are expected to do all the work, including non-horticultural work, that produces the range of social benefits we associate with urban cultivation, "out of care and love," and not as an occupation. Here, it is worth recalling that it was all the "extra work" that led to the growers' sense of burnout.

In recent years, funding opportunities for urban cultivation have expanded, with more philanthropic organizations, private investors, and governments showing interest in supporting the practice.[11] The USDA's Office of Urban Agriculture and Innovative Production has begun offering grants specific to urban growers,[12] with a particular emphasis on supporting "historically underserved farmers and ranchers."[13] This is a promising turn, particularly in light of the historical role of the federal government in encouraging and justifying policies that have dispossessed Black and indigenous people of their land and systematically undermined their capacity for food sovereignty.[14] The USDA and land-grant universities across the US have also expanded technical assistance to urban cultivation projects through agricultural extension services, some of them specifically set up in cities to serve urban growers. The operation and effectiveness of these programs merit ongoing scrutiny from scholars and activists alike.[15] In this context, recognizing the historical significance of what sociologist Bobby J. Smith, Jr. calls "food power politics," or the strategic use of food to mitigate social conflict, either in an oppressive or emancipatory way, becomes a significant framework for understanding why urban agriculture is now gaining prominence and who would stand to benefit from this trend.[16]

One particularly pressing question is the extent to which these new, well-intended forms of funding and technical assistance programs will create expanded opportunities for growers of marginalized backgrounds who harbor reasonable skepticism toward these institutions. Even when they do apply for external funding, the grants cycle pressures nonprofit growers to change their priorities or redesign their practices to match shifting foundation priorities. Funding for urban agriculture projects most often goes toward infrastructural costs, but what growers need support for most are day-to-day operational costs of sustaining existing projects, from rent, water, and labor to insurance and permitting. For-profit growers, for their part, must win the confidence of investors and banks who are more inclined to back business plans that fit existing business frameworks and use familiar terms that emphasize return on investment. Most potential investors expect projects to achieve financial self-sustainability within a year or two, but it often takes a year or two for growers just to set up or remediate their soil, work out seasonal adjustments for crop rotation, and establish a viable distribution system. Furthermore, Black, indigenous, and other people of color, as well as all women, frequently encounter discrimination from lending institutions.[17] This study illustrates that even growers with their own resources may face multiple challenges in launching and operating urban cultivation projects because so many aspects of urban cultivation remain outside of existing bureaucratic, social, and legal norms. These challenges are especially acute in historically marginalized communities that have faced multitudes of social injustices in their right to live and work in the city as a result of redlining, gentrification, and heir's property land dispossession.[18]

Regardless of the form of cultivation growers pursue, land insecurity is their most significant hurdle. This is especially the case for growers tending land that they do not own. The intensified pace of gentrification following the global COVID-19 pandemic threatens even the gardens and farms that previously had not worried about losing their leases. Without consistent access to land and resources, the long-term viability of urban cultivation remains in question, even if growers are dedicated and the community and consumers want to see cultivation projects continue. Community gardeners in many cities have successfully organized themselves to protect their practices by securing long-term land agreements underwritten by municipal government agencies[19] or through non-government organizations.[20] Yet even community gardens continue to face multiple threats from development. When the local government decides to prioritize the developmental interests in the land, community gardens are often pitted against affordable housing, while developers are rarely expected to compete on the same playing field.[21] Gentrification could create internal conflict resulting from changing grower demographics, in which long-term community gardeners' use of the land is often deemed "less desirable" or likely to devalue properties if it does not fit the taste and visions of what urban agriculture should be in the eyes of the government, neighboring businesses, and the public.[22] Many land access policies for urban agriculture explicitly consider it a cost-effective tool for *temporary* land management, rather than a form of permanent land use with its own merit.[23] How such dynamics affect the collective viability of urban cultivation like the projects that are focus of this book needs further investigation, with a particular focus on who is gaining and retaining access to space, what types of cultivation they practice, and where in the city they situate that practice.

One model for establishing land security for urban cultivation that has been gaining interest among growers and their advocates is the community land trust (CLT). In this collective ownership model, a nonprofit entity holds the title to the land and offers ground leases to those who use the land for purposes approved by the nonprofit entity's board.[24] It is commonly associated with affordable housing as an attempt to curtail the impacts of gentrification on low-income residents by restricting rent hikes or home-price increases on structures built on the land in trust. The CLT model offers a promising vehicle to provide long-term land tenure for urban cultivation practices.[25] The potential and capacity for the types of urban cultivation that emerge as prefigurative urbanism practices to develop into a more formal, long-term, collective entity such as a CLT remains to be seen. Scholars caution that establishing and managing a CLT requires multi-level coordination across grassroots organizations, community associations, and multiple government agencies, often mediated by nonprofit organizations representing various constituencies and interests.[26] These challenges may be particularly acute for urban cultivation, given its low potential for financial gain compared to other land uses like residential or commercial development.[27] Establishing a CLT requires a complex, long-term strategy and significant negotiations,[28] something that not all growers may be interested in doing. Cases of successful incorporation of urban agricultural land use in CLT, such as the Dudley Street Neighborhood Initiative in Boston, do indicate that such a collaborative model could work. Given their diverse aspirations and adaptability, some growers engaging in prefigurative urbanism may be open to this form of alternative model for gaining more long-term sustainability in their cultivation projects. But not all CLTs will view urban cultivation as fitting with their mission, nor would all cultivation projects seek to be a part of a CLT, as suggested by the heterogeneity and autonomy of the urban growers described in this book.

Cultivating Hope in Transitional Cities

Prefigurative urbanism figures largely in the increasing scholarship on unauthorized and informal uses of urban spaces, such as DIY urbanism and everyday urbanism. Some overlaps and ambiguity across these concepts are perhaps inevitable, considering that they have been developed to understand a set of liminal, and often hyperlocal, practices.[29] But whatever we call these practices, it is clear that some individuals are attempting to bring about change by challenging the norms of life in the city, whether through the physical transformation of space or by developing novel forms of social interaction. Activists and growth coalitions should also take note of these individuals and their everyday praxis, as the growers' preference to engage in prefigurative urbanism represents a level of distrust in both realms of social changes by some urbanites. These individuals are neither free-riders nor joiners, and they are not complicit in accepting the way the city is designed and managed by those in the positions of power. Actors who adopt practices of prefigurative urbanism operate in a distinct register from optimism, denial, or despair. They exhibit a pragmatic, willful hope through their individualized, immediate, and tangible actions.

But this book is not an endorsement or a celebration of these actions. Before anyone steps outside their door in search of a space to start a garden to enact their willful hope, I must remind them that, while some of the decisions and actions by the growers in post-Katrina New Orleans serve as inspiration, others offer cautionary tales. Because two of the principal characteristics of prefigurative urbanism are heterogeneity and ends-effacing openness, it would be ill-advised to create a manual for how to do it right. But there are nevertheless a few collective lessons from the experiences of the growers analyzed in this book that may better prepare someone who wishes to start an urban farm, a school, a business, or anything else as a form of prefigurative urbanism.

Understanding a place's history and social relationships, especially if you are not from the community, but even if you are, will help you better situate the practice within local understandings and expectations for what you are presenting as an *alternative*. However new and innovative an idea is to you, someone likely has already considered or tried a version of it. Find them and learn from them, or work alongside them if that is possible. It is also necessary, though it may be hard to do so when you are driven to enact your willful hope, to ask yourself if it is what people want to see happen in their community. Look for those who may resist or reject your ideas, and consider why they may disagree with your vision. Maybe it stems from misunderstandings or a lack of trust, but they may anticipate negative outcomes that you do not foresee based on their previous experience or knowledge about the place's history. Be aware that something that exceeds the bounds of existing legal and social norms can produce bureaucratic and other forms of consequences whose resolution may absorb a significant amount of time, energy, and resources. Being reflexive of your own social positionalities, including both privileges and limitations, will keep the alignment of your aspirations and practice in check as you continue to experiment and adapt over time. But no matter how much you prepare, you will probably make mistakes and adjustments as you go, and your actions will likely have both positive and negative consequences, including some unintended. Finally, be conscious of your own capacity—financially, physically, socially, and psychologically. It will change over time as you grow older, the circumstances around you shift, and your own knowledge and ideas about the practice evolve. Recognizing and communicating where you are in your prefigurative journey to the people you are working with allows all of you to plan better for a transition, when, or if, the time comes for you to step aside or away. The growers I studied had to learn these lessons on the job, and had to keep moving forward as they observed and reflected, because it was the *doing* that motivated them to continue enacting their willful hope.

∾

In 2010, artist Candy Chang placed thousands of stickers on vacant or blighted properties across New Orleans as part of a public art installation designed to engage the residents in conversation about their visions for the spaces' potential.[30] The red and white stickers looked like the ones typically used in social gatherings, with the phrase "HELLO MY NAME IS" replaced with the phrase, "I WISH THIS WAS." Members of the public were encouraged to fill in the blank space with their own wishes. Residents and passersby placed the stickers on the walls, fences, and posts to share what they wished to see, from types of businesses or services (e.g., grocery stores, flower shops, community centers, and community gardens) to poetic commentary on their sentiments regarding the state of the city or a particular neighborhood (e.g., "A higher priority in the city," "owned by someone who cared," "your dream"). The public art installation, described by some scholars as a form of DIY urbanism,[31] encouraged everyday urbanites, rather than planners, politicians, or business owners, to articulate their own imagination of what *could be*.

This book is about those who did not just wish for these changes, but actually made them happen. Did they all deliver on their aspirations? Not necessarily. While some projects became an impressive success, others struggled. Many mistakes were made. Growers might have done many things differently, had they known what was coming. But that was the point: No one knew what was coming. In the midst of the unprecedented, unpredictable, and rapid changes in the post-disaster city, where politicians schemed, businesses profited, and activists organized while outsiders openly debated whether it was worth rebuilding it at all, the growers cultivated the land. To cultivate land means to prepare it for growing plants. These growers built and improved the soil for their plants, but they were doing much more through their everyday practice of prefiguring an alternative—they were cultivating hope.

ACKNOWLEDGMENTS

I must first extend my gratitude to the individuals whom I got to interview and learn from for this project. They have been gracious enough to let me take up their time with interviews, emails for multiple follow-ups, and requests for site visits. Many of them even read portions of the book where they appear and trusted me to use their actual names and their stories. I could not include every aspect of each of their various accomplishments and journeys, but I have always admired the time and energy they dedicated to their projects, as well as their humility and candor. Without their generosity this book would not exist. I only hope that I do justice in accurately presenting and analyzing some key aspects of their collective experiences. I would like to especially recognize Pamela Broom, who gracefully granted me time to learn the history of urban cultivation in New Orleans, and also agreed to coauthor a book chapter and an article with me over the years. I look to her tenacity, kindness, and optimism as my North Star as I continue to learn.

I held positions at Tulane University and Georgetown University while working on this project, and I got a chance to work with some brilliant research assistants over the decade at both universities. Cate Irvin and Scarlett Andrews conducted two-thirds of the original interviews, and they were essential in identifying and recruiting the interviewees. They also helped digitally archive data, including old Parkway Partners files, and collaborated with me on a journal article based on the preliminary data analysis from the project. Mary Soule, Sarah Sklaw, Annie Mellan, Charles Miller, and Irene Chun assisted me in data collection, organization, and analysis. Theresa Werick was terrific in helping me sort through a clutter of data I had accumulated over the decade. I thank my colleagues at Tulane University Sociology Department and Georgetown University Sociology Department, where I got to present my earlier findings and receive constructive feedback. Corey Fields, Becky Hsu, Brian McCabe, Kristin Perkins, Kathleen Guidroz, and

Rahsaan Mahadeo have generously read and commented on the earlier drafts more than once, and I must also recognize their support for getting me back on the tenure track so I can have more time and resources to finish this book. The leadership and guidance of Michele Adams and Karolyn Tyson at each department were essential at the crucial moments in my career.

There are many others who supported this work along the way. I truly appreciate that Ilene Kalish of NYU Press saw potential in this project when many others passed because "there have been too many Katrina books." She not only kept me on track, helped me see the project's broader relevance, and got me through the finish line, but also thought it was worth considering my crazy idea to ask Candy Chang for permission to use images of her "I Wish This Was" public art project for the book cover (I still cannot believe she said yes!). I thank Priyanka Ray for her patience and always promptly responding to my many inquiries. Audra Wolfe of the Outside Reader provided extensive content editing expertise to help me gather my meandering thoughts and writings into a clear narrative that tells compelling stories. Line-editing by Matthew Somoroff significantly improved the manuscript by trimming the fat, ensuring consistency and flow, and dealing with my decades-long battle with misplaced articles in English writing. Carole Sargent created a space for me to work during my early years of exploring how to turn this project into a book through her scholarly writing group at Georgetown University and offered advice on the publishing process.

There are many folks who saw me struggled through this project over the years and guided me through the journey. I benefited from conversations with colleagues who shared encouragement and critical insights on my project throughout the process, including Justin Myers, Jeanne Firth, Joshua Sbicca, Deana Rohlinger, Paul Draus, Leslie Bunnage, Hamil Pearsal, Tina Rosan, Lindsay Campbell, Sanjay Kharod, Katie Acosta, Danielle Rudes, David Ortiz, Stephanie Arnett, Pepper Roussel, Elizabeth Chiarello, Kara Young Ponder, Randall Amster, and Devin Wright. I am also grateful to have been a part of the Food Justice Scholar-Activist/ Activist-Scholar Community of Practice, even for a brief period, in my development as a food justice scholar. I am especially indebted to Alison Alkon, who has always been a great mentor, and truly outdid herself by reading the whole draft with such short notice and providing sharp and constructive feedback, and then giving me the courage to release it into the world. David Snow and Calvin Morrill both have always responded whenever I reached out for support—nevermind that their formal obligation to advise me ended more than a decade ago. Dave and Cal, I try to pass your generosity forward in my own work. The urban growers and supporters in DC, especially Kate Lucy Lee and Joshua Singer, helped me see the book's generalizability beyond New Orleans.

Earlier analyses of data from this project have been published in *Urban Affairs Review* and *City & Community*. Portions of the data were included in my collaboration with Catarina Passidomo and Daina Cheyenne Harvey, which was published in *Local Environment* and *Urban Studies*. Some of the ideas for this book came out of my presentations at annual conferences of the Association of American Geographers, the Urban Affairs Association, the American Sociological Association, the Association for the Study of Food and Society, the Agriculture, Food and Human Values, Southern Sociological Society, and Humanist Sociology, as well as at the Green Cities: Inequality, Space, and Sustainability conference at Princeton University, the From the Outside In: Sustainable Futures for Global Cities and Suburbs conference at Hofstra University, and the Northeast Region Urban Agriculture and Sustainability Meeting at University of District of Columbia. The project benefited from funding from the two universities where I held positions, including a Murphy Institute grant, Community-based Research grant, Monroe Fellowship, Summer Research Fellowship, and Research Fellowship by Committee on Research at Tulane University, as well as Grant-in-Aid from Georgetown University, in addition to the start-up research funds I received from both institutions when I joined the faculty. The funding allowed me to provide compensation for the interviewees, travel for data collection and conference presentations, hire research assistants, and pay for professional transcription and editorial services, among other expenses related to this project.

My family's move to Washington, DC, in 2015 posed a new challenge in my journey with this project, not just because of the disruption in data collection continuity but also because of the loss of the community of support we built in New Orleans. While we miss all the unique aspects of the city, from hosting the Muses Thursday party during Mardi Gras to boiling sacks of crawfish in our backyard for all our friends and their friends, it is the people who became our extended family there whom I missed most intensely. Jill, J.J., Julie, and Brad, that corner of Lyons Street is forever the best intersection in the city, even if none of us live there anymore. Chrissy, Carter, Renata, and Jordan, thank you for always being there to remind us to enjoy life and be generous. Mary and Noriko, I am grateful that you always responded and made time for me whenever I showed up in the city with very little notice. Heather, little did I know I would end up consulting you multiple times on archival research when we first met at Abeona House. But beyond your support for my research, our frequent text messages about kids, NOLA, and politics kept my perspective in focus whenever I felt lost or ungrounded. Brack and Krista, no other restaurants but Cowbell would let our kids roam free and greet Uncle Scott in the kitchen while we enjoyed the best sangria and burgers in town. I miss that quirky spot so much, like everyone else in the city. Vaiden and Kaya, thank you for watching each of our children during their early months. It is no wonder they grew up to love arts and music. I did eventually build new communities in DC. Maura, Emily, and Paty, I feel extremely fortunate to have found you and your families. Our text chain, dinners, and camping trips have given me joyful moments of levity and reality-checks as we each went through various challenges of parenting and mid-life chaos, especially during the pandemic.

I would have dedicated this book to my mother, Yasuko, who passed away in 2022, just as I was starting to really dive into writing the first few chapters. But being a truly humble pragmatist, she would not have approved of such a grand gesture, regardless of the influence she had on me and for planting the seeds of this project decades earlier when she decided to turn our suburban backyard in Japan into a cultivation space. Because of her, I grew up eating organic, seasonal produce and did not even think to ask why our backyard was filled with rows of potatoes, tomatoes, onions, and citrus and fig orchards, rather than ornamental shrubs or stone gardens like our neighbors'. Though little Yuki did not help her parents in the garden very often, the idea of growing food in the city was instilled in me at that house, in close association with concerns about environment and social injustice, two issues that motivated our mother to do many things that did not fit the cultural norms of Japan at the time. I recognize in retrospect that she was engaging in a form of prefigurative urbanism. As her health declined, my father, Naoki, and my two siblings, Aki and Yutaka, assured me that I could stay focused on my work and my family while living half a world away from them. We continue to keep Okaasan's memory alive by talking about what Otousan's growing in the garden, and about my own tiny garden in DC, where I struggle to grow a few summer vegetables and fuss over how to prune the fig tree each year.

Finally, and most importantly, I thank my partner Keith and our kids Akira and Saya for their constant support. Keith, thanks to your sense of humor, we never take ourselves too seriously at our home, and your wide-ranging curiosity and ideas have taken us on so many memorable journeys that provided much-needed distractions whenever I was over-thinking or losing direction. I leaned on you heavily especially during the last year of completing this book, as you took the kids to their or-chestra rehearsals, birthday parties, and on many small and big trips so I could focus on writing. I know that I talked your ears off every time I had some roadblock or breakthrough with this project for over a decade, but you just let me hash it out, then encouraged me to keep at it. Thank you for your patience and I can't wait to see where our journey takes us next. Kids, one of you was in the sling when I started this project and the other was barely an infant as I wrapped up the initial data collec-tion. Each of you has accompanied me on some of the site visits, even if you do not remember, and you had to put up with me staring at my computer screen many weekends and evenings away from you. Now you see the book that I have been talking about most of your lives is finally here—and yes, of course you can read it, but I wouldn't be offended if you only read this part. Just know that I love you no matter what.

METHODOLOGICAL APPENDIX

If I could start this project over, there are so many things that I would probably do differently. But that is often not how social science research works, especially qualitative research that is based on particular people in a particular place at a particular moment. There is no do-over. The research for this book evolved truly in an inductive fashion, meaning that I did not start with specific hypotheses but rather that the empirical data guided methods and generated theoretical questions. Research questions and the project foci changed over time, and each time I reframed them I needed a different set of data or theories to make sense of what was happening and how it was happening. Everything was changing so quickly in New Orleans during the decade following Hurricane Katrina, economically, socially, politically, and culturally, and all of us were exploring, adapting, and learning as we went, including the growers and myself as a researcher. Because of this, I am certain that there are aspects of the topic that I failed to notice or look for in more than a decade of working on the project, and I take full responsibility for these oversights. Writing this book has taught me a simple yet valuable lesson on research, and possibly on life: in hindsight many things seem obvious, but they are not so when you are living through them. I provide more details on how this project came about so readers can gain a better understanding of its evolution and the decisions I made along the way as a researcher.

This research project started in its earlier conception around 2010, shortly after I took a position as a faculty member at Tulane University. Seeing the development of alternative food movements in the city, I conducted a series of preliminary research projects, mostly through Hollygrove Market and Farm, to understand who was engaging with this movement in the city, and more importantly, who was not. My focus on urban cultivation developed over time as I noticed the increasing number of gardens and farms across the city, and how they were gaining public attention. Around 2013, I developed the plan for a study that became the basis of this book with a focus on land access and tenure for urban agriculture, as I suspected that encroaching gentrification in the city was about to change the practice. My research primarily used in-depth, qualitative interviews of the growers and the representatives of the organizations that regulate or support urban cultivation in the city, but my analysis also included ethnographic observations, archival data, and media reporting on past and present urban cultivation practices, as well as geographic information on the cultivation sites. My relocation from New Orleans to Washington, DC, in 2015 marked the initial conclusion of the data collection. I later decided to conduct follow-up interviews from 2018 through 2023, making it a longitudinal study with an even longer time span (2005–2023) than had originally been planned (2005–2015).

The initial interview for this project took place between 2014 and 2015. I decided early on to exclude conventional community gardens from the study, because there was already a robust scholarly literature on this type of urban agriculture. In contrast, there was almost no systematic or longitudinal scholarship focused exclusively on what would typically be called "urban farms" at the time, especially scholarship that studied city-wide trends and trajectories rather than a single project. I later developed the term "urban cultivation" to dissociate it from "farming" and "agriculture," as I note in the introduction, but during the data collection process, I used the term "urban agriculture," as it was a familiar phrase to many practitioners and researchers. Similarly, we often used the terms "gardening" and "farming" interchangeably in interviews, though we also made a point to ask each interviewee which terms they preferred to use to describe their space (garden or farm) or themselves (gardener or farmer). It was in this process that I became familiar with the term "grower," which most individuals used to refer to themselves in lieu of "farmer" or "gardener," and I have used it throughout the book.

The sampling logic for the interviewees was to maximize diversity rather than to have a representative sample, since there was no formal list of all the growers in the city at the time. I also was more interested in understanding the range of experiences across different types of cultivation practices, locations, and grower demographics. We used a combination of online searches and personal references to identify potential interviewees, and only one individual explicitly declined to be interviewed after being invited to participate in the study during this process, though there were at least three growers who did not respond at all to our initial invitation for the interview. Seven of the original interviewees were no longer growing in the city by the time of the initial interview period in 2014 and 2015, but had at some point started or worked on an urban cultivation project that began after 2005. For the follow-up interviews between 2018 and 2020, I reached out to those who had still been actively growing in 2015. Of the twenty-seven who fit that description, I was able to contact twenty-four. I conducted follow-up interviews with all but one of those twenty-four. One person was interviewed in 2018 for the first time, because they were originally not available during the initial interview period even though they were already operating a project in the city at the time. Follow-up interviews were not part of the original research plan, and thus do not have the structural consistency of what are called "serial interviews," often used in medical research for a long-term follow-up with patients, or the "longitudinal qualitative interview" in longitudinal, panel social science research that aims to follow the same individuals over time to understand their trajectories.[1]

Crucially, I did not interview any of the residents who were neighbors to the cultivation projects. Including their perspectives would have provided more nuanced, potentially conflicting, accounts of the impact that these gardens and farms were having in the community. This would have been especially meaningful in addressing the underlying tensions around racial, class, and generational divisions in these communities, to better understand where these growers, especially those who were white and not from New Orleans, were situated in the hierarchy of status and local social networks. Instead, I relied on the growers' description of their relationship with the neighbors, though I have tried to make it clear in my writing that this was how the growers saw it, rather than presenting such description as empirical fact.

The initial interviews, conducted by Cate Irvin, Scarlett Andrews, and myself, took place in various locations, based on the interviewees' preferences. Gardens and farms were mostly quiet and great for gaining insights about the space and the growers' cultivation practices, but nature caused some unexpected disruptions—gusts of wind, rain, heat, and bug bites. Coffee shops were cozy and more relaxed, but the loud sounds of a vibrant "third place" created a headache for the transcribers. On a few occasions we ended up conducting impromptu "going along" ethnographic interviews,[2] as we were unexpectedly taken along on grower errands or asked questions as they walked through the field to check on their crops. I conducted six of the follow-up interviews in person when I visited the city in 2018, and others over the phone or on Zoom. In addition to these formal interviews, I had irregular communication with ten growers over the course of this project, without audio recording.

The initial interviews focused on land access, but also included a set of questions about growers' motivations for engaging in cultivation projects, what they were growing, how they were growing, and what they were doing with what they grew. We ended each interview by asking growers about their plans for the next "five, ten years," and what they anticipated as the potential hurdles for achieving these goals. The follow-up interviews picked up where the initial interviews had left off. These interviewees were provided with the transcription of the original interviews and asked to assess how close their trajectories had been to the vision they expressed in the original interview. The follow-up interviews also included more questions about the growers' view of the development of the urban cultivation scene overall, beyond the interviewee's own cultivation project. In 2023 I met with eleven of the original interviewees who were still in New Orleans at least part-time; I wanted to learn about where they had been, including during the pandemic, and share my ideas for the book. All interviews except for the three follow-up interviews and 2023 conversations were digitally recorded and transcribed word-for-word.

Ethnographic observation was not a primary method of data collection for this project, but between 2012 and 2014 two undergraduate students and I volunteered at Hollygrove Market and Farm, and another undergraduate student volunteered at Our School at Blair Grocery, and conducted participant observation. I also participated sporadically in events held by various cultivation projects, fieldnotes from some of which have been included in the book. While I wished I could have spent more time conducting field research for the project, the responsibility of being an assistant professor and having two small children during the initial data-collection period limited my capacity to engage in extended, embedded on-site observation. I did conduct regular check-ins on the cultivation projects several times a year by driving or walking by the projects to see if the cultivation sites were still in operation. I continued to visit sites during my half-dozen trips to the city for conferences or vacations between 2015 and 2023, and noted which gardens were still in operation, which no longer seemed active, and where new gardens and farms were being set up. These informal visits and observations also gave me a sense of how the neighborhoods around the sites were changing, too. My deepest regret is that I did not take photographs of many of these projects when they were still in operation, especially the ones that seemed to be thriving in 2015. Little did I know that so many of them would cease to exist by 2018. Quality of cameras on smartphones has significantly improved since 2014, and if it was available then it probably would have occurred to me more often to capture the scenes of action or just the aesthetics of the sites in the context of the community in a particular time and place. In some cases I could virtually revisit the site during a particular month and a year on Google Streetview, but so many of the cultivation sites were on streets or alleys less frequently captured by the service.

As I conducted interviews and began to educate myself on the vast scholarship of environmental justice and food justice, I realized that the post-disaster trajectory of New Orleans's urban cultivation scene had to be understood in the context of the city's history. I began to hear stories about the ubiquity of local food provisioning practices long before Hurricane Katrina, especially in the city's Black communities, from older long-term and multi-generational residents. This led me in search of documentation of the historical precedence of urban cultivation practice in the city. I conducted archival research at Tulane University's Louisiana Research Collection and identified some historical documentation of gardening practices by the early French colonists, German and Italian immigrants, and enslaved African people in the city. Heather Green suggested I look up the Notarial Archive, where I was able to peruse colorful renderings of vegetable gardens in Plan Book of Plans from centuries ago. Per a suggestion from Pamela Broom, I consulted the documentary *In Search of Yesterday's Gardens* and Jacques-Felix Lelievre's *New Louisiana Gardener* to gain insights into how urban cultivation in the city's earlier decades has been documented, especially by white New Orleanians and the city's European visitors. Vern Baxter at the University of New Orleans directed me to Tony Margavio's work on Italian immigrants in New Orleans, which includes their engagement in truck farming practice. But overall, formal documentation of communal urban cultivation practice in Black communities was notably absent from archival holdings, compared to the stories and reporting of such practice in immigrant communities, from the Italians and the Germans during the nineteenth century to the Vietnamese refugee community since the 1970s.

In addition to the historical archive, I continued to collect local and national media coverage of New Orleans's urban cultivation scene. I set up Google alerts with search terms such as "urban agriculture New Orleans" or "urban garden New Orleans" to catch reports or writing on relevant topics. I also read local news and environmental or food blogs including the *Times-Picayune* (nola.com), the *Advocate*, the *Lens*, *Gambit*, *Antigravity*, and *Scalawag* to keep abreast of emerging trends and public perceptions of urban cultivation practices in the city. The historical archival data and media coverage helped contextualize my data analysis, allowing for a broader understanding of the long-term trajectory of the city's development and its impact on who was growing and eating locally before the most recent emergence of urban cultivation practice.

Transcribed interviews and fieldnotes were coded using Atlas.ti qualitative data analysis software. I adopted the inductive coding strategy of Grounded Theory Method by creating and assigning codes to the contents of the interviews without pre-established hypotheses.[3] Over time, several themes emerged out of over 400 codes that I created: land access processes and outcomes; grower aspirations; obstacles and opportunities in establishing and sustaining cultivation projects; grower relationships with other growers and surrounding communities; the role of external actors such as government agencies and nonprofit organizations. I came to recognize that the dominant set of aspirations and practices changed over time, but that they also changed within each project. The expanded timeline of the study also enabled a deeper and more complex analysis of the relationship between the growers, the land, the communities where they were growing, and other growers, as I took note of why some of the projects persisted and others terminated over time.

The data collection and analysis methods used for this study were approved by the Internal Review Boards of Tulane University (11–192523U;

12–296658U; 11–290013U) and Georgetown University (CR00001874). The consent form for the initial interview specified that the interviewees' names would be kept anonymous but the projects would be named, and all interviewees agreed to this arrangement upon signing the consent form. However, I later realized that this would not provide sufficient confidentiality for the growers because an Internet search could easily identify these individuals if specific cultivation projects were named in publication. Thus, I provided options for those invited for the follow-up interviews to be named or given pseudonyms in the finalized publication in the consent form. At that point, all but one of them agreed to have their real name used in the book, though two individuals subsequently changed their minds and opted for pseudonyms. After the initial draft was completed, I shared the sections of the book where these individuals were mentioned, with or without direct quotes from interviews, and confirmed that they were still comfortable with being named. To be consistent with the original consent form, those who were not invited for follow-up interviews and others that did not confirm their willingness to be identified were either kept anonymous or given pseudonyms if their experiences or quotes appear more than once in the book. For various reasons there are many whose stories never made it to the book, but their experiences and perspectives were nevertheless part of my analysis.

In the process of confirming whether or not the interviewees wish to be named, some details such as timeline and location were corrected. In just a few cases, the interviewees pointed out that my interpretation of the intent or nature of some events did not match their recollection (i.e., not necessarily that the event did not occur, but how they recalled it), which I honored. Notably, these were instances where I implied political intentions in my analysis where the actors felt there were none. Some of the interviewees asked if I would consider cleaning up their transcribed speech, because they felt that verbatim transcriptions made them sound much less articulate. My initial intention was to keep all of the "um's" and "like's" in their speech to recognize that these were statements of someone on a particular day when they were put on the spot to answer a question for a research project, and not necessarily the static and formulated ideas that they held constant. In fact, some of the interviewees noted that it was a "travel down memory lane" or even embarrassing to hear themselves from years before, especially if their practice or way of thinking had since evolved. Nevertheless, I came to realize that in some cases incomplete sentences or a discontinuous flow of speech made it difficult to follow their main arguments or sentiments. Thus, in the end I opted to "clean up" some of the speech whenever I was asked to by the interviewee or in consultation with the readers of earlier drafts, so that it is easier for the reader to follow the speaker's train of thought, without making them completely grammatically correct at all times.

As noted in the Introduction, I have striven to be reflexive about my "outsider" positionality in this project. Aside from the biographical differences between myself and the growers, I was also conscious of my affiliation with Tulane University, a private institution with a mixed local reputation, especially among low-income communities who remain skeptical of its contribution to the local economy and culture. Some of the growers expressed interest in wanting to work with "someone from Tulane," with an implied assumption of possible access to resources and legitimacy through affiliation, which I tried to make clear that I could not provide. Others expressed disdain or distrust of the university based on their previous experiences of working with a faculty member or students that left them feeling disrespected or taken advantage of. My general inclination was to distance myself from the university by reminding the participants that I was just an assistant professor who worked there. Partly as a form of keeping symbolic distance from the university, I chose not to use "service learning" as a part of my research despite its popularity on campus, especially after hearing about the problematic implementation that often benefited the students and not the cultivation projects.

Another source of hesitation or skepticism from the public toward researchers stemmed from the exploitative nature of studies that took place in the immediate aftermath of Katrina. A large number of researchers descended upon New Orleans, collected data on ecological, infrastructural, and psychological damages in the city, and then left, often without follow-up care for the secondary trauma caused by the data collection process. These researchers rarely shared the study results with the participants. These practices raised questions about the ethics of studying post-disaster communities.[4] I tried to minimize these harms in my data collection and analysis process by ensuring interviewees' voluntary participation and providing avenues of communication to clarify, modify, or redact what they had shared during an interview, and by engaging some of them in the research question and analysis development process throughout the project.

The most challenging aspect of inductive research is theorizing what you see in the data, to identify patterns, groups, and trends. I dragged this part out, partly due to some life events that made it difficult to find time to do the kind of deep thinking that is necessary to make sense of the messy qualitative data of fifty interviews, geographical information, and archival data. The process reminded me of the Magic Eye trick books that were popular when I was younger: pages were filled with wavy shapes in psychedelic colors that looked messy and nonsensical when you first glanced at them. But as you fixated your eyes on these pages, eventually shapes emerged that you did not see before, and once you saw them, they were so obvious. For more than five years, I read over these interviews, first to code them line by line, then to look through some themes using various analytical features of Atlas.ti, in addition to rereading the transcripts or listening to the original recorded interviews multiple times. I tried classifying growers according to different demographic attributes (e.g., age, gender, race, birthplace), their organizational structure (e.g., for-profit, nonprofit, volunteers, paid employees), or how and what they grew (e.g., commercial sales, donation, vegetables, flowers, animal husbandry). None of these proved to be theoretically meaningful categories at first, and the growers kept falling into multiple categories at once, defying my attempts to put them into discrete boxes to distinguish them, until I began to develop the aspiration typology described in Chapter Two.

I continued to immerse myself in the emerging and quickly expanding scholarly literature on food justice and environmental justice, especially work that dealt with urban cultivation. This deep dive into the scholarship was extremely helpful in showing how to situate urban gardens and farms in the context of the larger political economy of the city, and to understand the role it could serve as a place for political mobilization, especially in cities in the Global South. But this scholarship also made me insecure because I was not seeing these same patterns in my data. The growers in my study did not proclaim to be activists, and my analysis did not show that they were coalescing as a collective movement across the city. I even posed the question directly during the follow-up interviews and in my final check-in conversations with growers; all concurred that it was not a movement, even those who wished to see more collective mobilization. In short, my data did not fit the political gardening framework that seemed prevalent in the food and environmental justice scholarship on urban cultivation, which became an impetus for me to develop a new concept that explain why the existing theories did not apply to my data.

My reading of the vast body of scholarship of post-disaster New Orleans also left me without clear directions, as it was mostly focused on the tension between the neoliberal remaking of the city and the grassroots resistance against such forces. Again, my data failed to fit these existing theories, as the growers largely operated outside these two realms. Many started their projects with some vision for social change but were not part of the official post-disaster recovery agenda, as it was the case for education or housing reform. Many growers took advantage of the new food economy that was part of the redevelopment, but they did so selectively and did not see significant financial return on their participation. They were not part of nor did they emerge from the long-term forms of community activism that had existed in New Orleans. So, what were they? This is how I developed the theorization of the growers' practice as a form of prefigurative urbanism, by merging my understanding of prefigurative politics scholarship, urban studies, and cultural sociology, to find meaningful similarities and differences across cultivation projects and their trajectories over time. In the process, I also consulted a wide range of scholarship in disciplines beyond the social sciences, such as philosophy and history, in search of theories that were applicable to the data I collected. This is how I eventually settled on framing the growers' practice as a form of prefigurative urbanism that sought to activate hope in a post-disaster city.

Throughout the writing and editing process, I struggled to find the balance in my writing between being overly sympathetic toward the growers and simply reducing them to stereotypical saviors (especially the white, privileged post-Katrina transplants). There was a part of me that felt uneasy about not taking a stronger critical stance that would have aligned this book better with the dominant discourse within food justice scholarship, which leans decisively toward progressive, radical perspectives. Ultimately, my analysis and writing reflect the data I was able to gather for the book, and I recognize that another scholar could have approached this topic from a different vantage point and written a different book. As a researcher, you get used receiving critiques on your manuscript as a part of the routine peer-review process in publication or at conference presentations. But those moments when I shared my preliminary findings in conversation with the growers or sought their feedback on the sections of the book where they themselves appear— that was the most nervous I had ever been as a researcher. It was not so much that I wanted them to like what I wrote, but I wanted to make sure that I got it "right." I hope it at least comes close to that as a whole.

NOTES

INTRODUCTION

1 Lawson 2005; Bassett 1981.
2 Myers 2022; Sbicca 2018; White 2010.
3 Shostak 2021.
4 Hongagneu-Sotelo 2014; Shostak 2021.
5 For the history of redlining, residential segregation, and political disenfranchisement, refer to works by scholars such as Massey and Denson 1993; Rothstein 2017.
6 White 2010; Lindeman 2022; Lawson 2005.
7 Lawson 2005, Chapter 7; Haydu 2021, Chapter 4.
8 Tornaghi 2017; Reynolds and Cohen 2016.
9 Penniman 2018; White 2010.
10 Logan and Molotch 1987; Smith 1987.
11 Lawson 2005; Saldivar-Tanaka and Kransy 2004; Irazábal and Punja 2009.
12 Lawson 2005; Basset 1981; Pothukuchi 2017.
13 Schilling and Logan 2008.
14 There is an expansive scholarship on the causes, processes, and consequences of
   gentrification. For the most recent summaries of the state of the scholarship, see
   Brown-Saracino 2016; Zukin 2016; and Lees and Phillips 2018.
15 Monteiro et al. 2020.
16 Gould and Lewis 2016; Dooling 2009; Checker 2020.
17 Logan and Molotch 1987.
18 Dooling 2009.
19 Campbell's study of PlaNYC initiatives presents an exception here. See Campbell
   2017.
20 White 2010; Sbicca 2018; Broad 2016; Hassberg 2020.
21 Nettle 2014; Tornaghi and Dehaene 2020; Mason 2014; Holt-Giménez and Shattuck 2011.
22 Logan and Molotch 1987.
23 Gotham 2007; Adams and Sakakeeny 2019.
24 Bullard and Wright 2009; Dyson 2006; Johnson 2011.
25 Fussell, Sastry, and VanLandingham 2010.
26 Luft 2008, 2009; Flaherty 2010, 2016.
27 Gotham and Greenberg 2014.
28 Adams and Sakakeeny 2019.
29 Thomas 2009.

30 Gotham and Greenberg 2014.

31 Harvey 2007.

32 Boggs 1977, 2.

33 White 2011.

34 Swain 2019.

35 Hondagneu-Sotelo 2014; Irazábal and Punja 2009; Penniman 2018.

36 Bherer, Dufour, and Montambeault 2024.

37 Ocejo 2017.

38 McGeer 2004.

39 Ibid., 122.

40 Eagleton 2015, 85.

41 One person declined to be interviewed for follow-up and others could not be reached. Four of these interviews were conducted as a more informal check-in conversation without audio recording. There were ten growers with whom I communicated, without audio recording, a few additional times as a follow-up conversation in addition to the original and formal interviews. For more details on methodology, please see the appendix.

42 One of them identified as "French" which was coded as "white" in the context of the interview. Two of the individuals use the pronoun "they" and will be referred to as such in the text.

43 The original interview project, conducted 2014–2015, was approved by Tulane University's IRB Office (192523–6). The consent form specified that the cultivation projects would be identified by their actual names but that the interviewees themselves would remain anonymous in research publications. The follow-up interviews were approved by Georgetown University's IRB Office (2018–0140). The consent form for these interviews provided the participants with the option of being identified by their actual names, after reviewing their original interview transcription. The following names in this book are pseudonyms: Bill, Erin, Colleen, Nicole, Shannon, and Ashley.

1. AFTER THE RAIN

1 Langenhennig 2015.

2 Erikson 1976.

3 Prior to French colonization, the area was home to the Choctaw and Chitimacha peoples, who had established their own foodways that were based on gathering local fauna and fishing in the river and the bayou (Densmore 1944). This book is focused on the colonized and urbanized New Orleans, but it is imperative to acknowledge that the local foodways of the region predate colonial invasion of the Mississippi riverbend.

4 Gotham 2007.

5 Broom and Kato 2020.

6 Douglas 2011, 106.

7 Reeves 2001.

8 Palmer 2006, 189.

9 Though it is spelled "truck farm" and the contemporary version involved the use of pick-up trucks, the term "truck" originally derived from a French word "torquer" which means to exchange or barter, not the transportation vehicle. See Campanella 2020.

10 Campanella 2020; Margavio and Salomone 2014.

11 Peters 2020.

12 Brown 2018; Fussell 2007.

13 Paul R. Valteau Jr., interviewed by The HistoryMakers, 2010. (Transcription provided by the archive.)

14 Wilkerson 2010.

15 The departure of the Black population from the rural South was not by choice. The New Deal's Agricultural Adjustment Act and the Farm Security Administration actively facilitated removal of Black farm workers and sharecroppers by subsidizing crop production and evicting Black sharecroppers to make the land available for white farmers. See Adams and Gordon 2009; Rosenberg and Stucki 2019; Scott and Brown 2008, Chapter 8.

16 Lawson 2005, Chapter 4.

17 One report estimates the value of crops being grown in New Orleans at $125,000 around 1918. Lawson 2005, 141, originally reported in Pack 1919, 98.

18 Lawson 2005, Chapter 6.

19 Landphair 2007, Paragraph 15.

20 By one estimate, New Orleans had approximately 26,000 vacant lots in 2000. City of New Orleans 2016.

21 Peters 2020.

22 Broom, Kato, and Roussel 2024.

23 Broom, Kato, and Roussel 2024; Litwin 1981.

24 Jensen 1986.

25 Cherrie 2014.

26 Bankston 1998.

27 Airriess and Clawson 1994.

28 Pottharst 1995.

29 Personal communication with Kris Pottharst on June 19, 2015.

30 Parkway Partners record obtained, organized, and analyzed by the author in 2014.

31 Ibid.

32 Lawson 2005.

33 To date, there has not been a definitive scholarly study that attributes specific factors to the decline of the folk urban cultivation in urban Black communities.

34 Total Population, 2000, 2010, 2015. *Social Explorer* (based on data from US Census Bureau; accessed December 18, 2023).

35 The total number for some categories does not add up equally across years due to lack of sufficient data to validate the information from every grower.

36 This number only refers to the individuals growing food at scale, either at community gardens or on larger city lots, and therefore does not include community gardeners or home gardeners.

37 Picou 2009.

38 Campanella 2009.

39 Kotkin 2014.

40 Plyer and Ortiz 2012.

41 Author calculation based on the 2000 U.S. Census (Table H004).

42 Fussell, Sastry, and VanLandingham 2010.

43 The Louisiana Road Home program was launched in June 2006, funded by the US Congress. It offered three financial compensation options for the homeowners whose properties were damaged by Hurricanes Katrina and Rita: a one-time grant to rebuild, the purchase of another home in the state, or selling their home and relocate outside of Louisiana. But these options were not really *optional* for many residents for various reasons. The value of the grant was assessed using pre-Katrina estimates of property values, rather than the actual cost of rebuilding homes. This meant that people who owned homes in the areas where the properties had been depreciating, who were also much more likely to have seen more significant damage to their property due to flood water distribution, had to come up with their own funds to complete the rebuild. For more on the racial disparity implications of the Road Home program, see Gotham 2014.

44 Article 880, Intestate Succession. Acts 1981, No. 919, §1, eff. Jan. 1, 1981.

45 Kluckow 2014.

46 Tremé Season 1, Episode 2 (Meet De Boys on the Battlefront).

47 Rich 2012.

48 Adams 2013; Harvey, Kato, and Passidomo 2016.

49 Ehrenfeucht and Nelson 2013; Weil 2008.

50 Author calculation based on Parkway Partners records.

51 Gotham 2007.

52 In 2018, Make It Right was cited in a class action suit for poorly constructed homes that did not hold up to the climate and usage over time, and for the organization's failure to respond and address concerns expressed by the owners.

53 Gotham 2017.

54 Rose et al. 2009.

55 Kato 2013.

56 In 2016, the City Center was renamed the Small Center.

57 Author's personal communication with a NORA staff member in 2014.

58 City of New Orleans 2011.

59 Vickery 2014.

60 New Orleans Strategic Hospitality Task Force 2010.

61 Gotham and Greenberg 2014.

62 Mayer and Goldman 2010.

63 Schor 2020.

64 Greenberg 2019.

65 U.S. Census Bureau 2015.

66 Rich 2012.

67 Mwendo 2012.

68 Author's personal communication with NORA staff in 2014.

69 Alkon, Sbicca, and Kato 2020.

70 Alexander 2019; Warner 2001.

71 Irvin 2017.

72 Firth 2023.

73 The Eat Local Challenge New Orleans 2015.

74 Moskowitz 2015.

2. SEEDS OF HOPE

1 Tierney and Oliver-Smith 2012; Erikson 1976.

2 D'Andrade 1992.

3 Swidler 2003.

4 Swidler 2003.

5 Gould and Lewis 2016.

6 See Allen 2008; DeLind 2011; DuPuis, and Goodman 2005; Guthman 2008.

7 The "food desert" concept was gaining public recognition across the US around 2010, partly reflecting the increasing public concerns that food consumption was a significant factor in the concentration of chronic illness and childhood obesity among low-income communities and communities of color. Scholars and activists increasingly use the term "food apartheid," coined by the activist Karen Washington (Brones 2018), to describe the structural and intentional racial injustice that caused disparate food security experiences along racial and class lines.

8 Adams 2013.

9 Flaherty 2016.

10 Cannon 2014.

11 Snow and Benford 1988.

12 *Reversing the Mississippi* 2015.

13 Dees 1998; Peredo and McLean 2006.

14 The initial set of interviews took place few years after Propeller, a local "social venture" co-working space and incubator, launched their first accelerator program in 2011.

15 Ocejo 2017; Mayes 2014.

16 For the overview of how social movement scholars use the concepts of framing, primarily in studying social movement organizations, see Snow et al. 2019.

3. BREAKING GROUND

1 McGeer 2004.

2 Rabito et al. 2012.

3 Egendorf et al. 2020.

4 There were inconsistencies in the growers' understanding of whether or not this was a legal requirement or just "good practice," especially during the recovery and the transitional periods. By the end of the decade, the requirement had been clarified, and most growers knew that they did not need to have the expensive backflow preventer installed by a specialist and that it was sufficient to use another tool called a backing breaker, which performed a similar function but cost significantly less and did not require specialist installation.

5 Article 880. Intestate succession. Acts 1981, No. 919, §1, eff. Jan. 1, 1982.

6 Historically, the succession law has disproportionately affected Black homeowners negatively, leading to the loss or devaluation of their properties. See Gibson 2022.

7 Hou 2010.

8 Kato, Andrews, and Irvin 2018.

9 Author's personal communication with Johanna Gillian in 2015; Firth 2023, 114.

10 New Orleans Redevelopment Authority n.d.

11 New Orleans Redevelopment Authority 2014.

12 The mineral rights clause was not the agency's own requirement but was a result of the property's acquisition by the Road Home Corporation. One of the NORA representatives later explained via email that "Road Home Corporation has reserved all mineral rights unto itself and NORA is incapable of conveying them to a subsequent purchaser."

13 Kato, Andrews, and Irvin 2018.

14 Propeller 2012.

15 McClintock 2010.

4. GROWING CHANGES IN A GROWING CITY

1 Gessler 2020.

2 Ocejo 2017.

3 Adams 2013.

4 The executive director of Market Umbrella, the organization that manages Crescent City Farmers Market, confirmed during an interview for the study that urban farmers do face these constraints when participating in their weekly market.

5 Section 4.2. RS-1A Single-Family Residential District. Permitted Uses.

6 Ibid.

7 Broom and Kato 2020.

8 Geographer Catalina Passidomo's ethnography of the Lower Ninth Ward documents how Backyard Gardeners Network, led by Jenga, "mobilized residents—many of whom had connected through events and activities at the Guerrilla Garden—in an eight-month long collaborative process to draft a 'food action plan' for the neighborhood." See Passidomo 2016, 15.

9 Guthman 2008.

10 U.S. Census Bureau QuickFacts: New Orleans city, Louisiana n.d.

11 Firth 2023, Chapters 4 and 5.

12 Shostak 2022.

13 Oldenburg 1989.
14 Erikson 1976.
15 Weems et al. 2010.
16 Jacobs 1961.
17 Ibid.
18 Granovetter 1973.
19 Swain 2018.

5. WHEN THE GARDEN GOES FALLOW

1 Taylor 1989.
2 Corriall-Brown 2011; Bunnage 2014.
3 Polletta and Jasper 2001; McAdam 1986; Rupp and Taylor 1987.
4 Lueck 2007.
5 McGeer 2004.
6 Irazábal and Punja 2009.
7 Glennie 2020.
8 Willson 2018.
9 Some of these issues had been identified prior to the walkout by Ceaser 2012.
10 The letter was obtained from a former staff member via email shortly after it was sent to the directors.
11 The documentary *Reversing the Mississippi* chronicles a part of this new direction, including attempts to invent new farming tools in collaboration with an open source hardware advocate, Marcin Jakubowski.
12 Ceaser 2012.
13 Adams 2013.
14 Ceasar 2012.
15 Corrigall-Brown 2011; Bunnage 2014.
16 Gorski, Lopresti-Goodman, and Rising 2019.
17 Gorski 2019a.
18 Chen and Gorski 2015.
19 McGeer 2004.
20 Gorski 2019b.
21 The Lower Ninth Ward Urban Farming Coalition, n.d.

6. CULTIVATING HOPE IN A POST-DISASTER CITY

1 For the neoliberal critiques, see Brand and Baster 2020; Johnson 2011; Bullard and Wright 2009; Hartman and Squires 2006. For grassroots responses, see Flaherty 2010; Luft 2008.
2 Putnam 2000.
3 Alvarez 2008.
4 Jordan and Maloney 2006; Hensby 2021.
5 Pew Research Center 2023.
6 Haenfler et al. 2012.

7 Adams and Raisboough 2010; Johnston et al. 2011; Schoolman 2019.
8 For example, see Johnston and Bauman's work on foodies' contradictory embrace of democracy and inclusivity while underplaying class privilege. Johnston and Bauman 2010.
9 Cho 2006; Dacin et al. 2011.
10 INCITE! 2017.
11 Finley and Esposito 2012.
12 Olson 1971.
13 McAdam 1986.
14 Taylor 1989.
15 Jeffrey and Dyson 2021, 44.
16 Swain 2019.
17 Törnburg 2021.
18 Finn and Douglas 2019, 20. For the remainder of the book, I use "DIY urbanism" to refer to these phenomena in general.
19 Pagano 2013; Douglas 2014; Volont 2019.
20 LaFrambois 2017.
21 Bherer, Dufour, Montambeault 2023.
22 For framing of citizens as "consumers" in the neoliberal disaster agenda, see Adams 2012, 195; Gotham 2012, 635.
23 Tierney 2015.
24 Pottinger 2017.
25 Swidler 2003.
26 Freudenburg et al. 2009; Gotham and Greenberg 2014; Brunsma et al. 2007; Horowitz 2020.
27 Gotham 2012; Johnson 2011; Adams 2013.
28 Klein 2008.
29 Gotham and Greenberg 2014.
30 Adams 2012; Fussell, Sastry, and VanLandingham 2010; Gotham 2015.
31 Akers 2012; Klein 2008; Cannon 2014.
32 McKeever and Pollak 2015; Firth 2023.
33 Adams 2013.
34 Clark 2010.
35 Arena 2012.
36 Harvey, Kato, and Passidomo 2016.
37 Luft 2008, 2009; Flaherty 2016; Leong et al. 2007.
38 Flaherty 2010; Tang 2011.
39 Tierney 2015.
40 Johnson 2015; van Holm and Wyczalkowski 2018.
41 Fussell 2015.
42 Campanella 2013.
43 Zukin 1982.
44 Glennie 2020; Irazábal and Punja 2009; Von Hassell 2002.

45 Fussell, Sastry, and VanLandingham 2010

46 Minkoff-Zern 2014.

47 Mayes 2014; Smith 2021.

48 Erikson 1976; Gill 2007.

49 Flaherty 2016; Keller 2022.

50 Dooling 2009; Gould and Lewis 2016.

51 McClintock 2014; Bherer et al. 2024.

52 Campbell 2017, 196.

53 Swidler 2003.

54 Adams 2013.

55 Swidler 2003.

56 Flaherty 2016; Firth 2023.

57 Ehrenfeucht and Nelson 2013.

58 Florida 2002. Creative class theory has been extensively critiqued, for example, see Peck 2005.

59 Campbell 2011.

60 For example, Our School at Blair Grocery had a penchant for prominently using these activist phrases in their programming and public engagement, but those involved with the organization questioned the extent to which the organization "walked the walk," as evidenced by the staff walkout in 2013, the documentary film *Reversing the Mississippi*, and a study by Ceaser 2012. Grow Dat Youth Farm, by contrast, has integrated these concepts into its curriculum and staffing, while retaining its focus on youth leadership training.

61 White 2010; Rosan and Pearsall 2017; Montgomery 2020; Cornelissen forthcoming.

62 Broom, Kato, and Roussel 2024.

63 Draus 2009.

64 Bankston 1998.

65 Vietnamese American communities in the Southeast Louisiana region were significantly affected economically by the Deepwater Horizon Oil Spill in the Gulf of Mexico in 2010 due to its impact on the seafood industry, in which the majority of the community members worked in some capacity. See Patel et al. 2018.

66 Jeffrey and Dyson 2021, 651–652.

67 Ibid., 652–653.

68 Benford and Snow 2000.

69 McMillan Cottom 2020.

70 McMillan Cottom 2021.

71 Eagleton 2015.

72 What I refer to as "pragmatic willful hope" here is to be distinguished from the "pragmatic optimism" concept that is most often associated with corporate leadership on sustaining organizational operation during times of crisis by balancing an optimistic outlook (wishful hope) with recognition of the flaws in the situation. The pragmatism in such rhetoric does not require a fundamental rejection of the

existing system, but merely encourages the actors to recognize its shortcomings toward fulfilling their pre-set goals (profit maximization), thus fundamentally distinct from prefigurative urbanism.

73 Checker 2020.

74 Reynolds and Cohen 2016.

75 Some examples of this type of urban cultivation projects are East New York Farms!, Detroit Black Community Food Security Network, and Community Services Unlimited in Los Angeles. See Myers 2022; White 2011, 2010; Broad 2016; Hassberg 2020.

CONCLUSION

1 Solitary Gardens n.d.

2 National Young Farmers Coalition, 2018.

3 Born and Purcell 2006; DuPuis and Goodman 2005; DeLind 2011; Hinrichs 2000.

4 Sbicca 2019.

5 See Campbell 2017.

6 Silverman et al. 2023.

7 Despommier 2020.

8 Benke and Tomkins 2017; Engler and Karrt 2021.

9 Wittman, Dennis, and Pritchard 2017.

10 An exception is the media coverage associated with the dissolution of Growing Power in 2017. Growing Power began as an effort to address food insecurity in low-income neighborhoods in Milwaukee, Wisconsin, in the 1990s, but its operation expanded rapidly after founder Will Allen won a MacArthur "Genius" Award in 2008. Its demise has been attributed to a combination of financial and organizational mismanagement in the context of growth, as the organization ballooned to accommodate demands for training and media exposure. Still, many consider its domestic and global legacy immeasurable.

11 Library of Congress Research Guides n.d.

12 US Department of Agriculture Urban Growers n.d.

13 The Agriculture Improvement Act of 2018 (2018 Farm Bill), which established the Office of Urban Agriculture and Innovative Production, also included provisions that identify "historically underserved" farmers and ranchers. The USDA provides funding and outreach to meet the unique needs of these growers in urban and rural areas.

14 Broom, Kato, and Roussel 2024.

15 Several of these urban agricultural policies specifically focus on cannabis. Given the historical impact of marijuana criminalization on communities of color, resulting in mass incarceration and an informal economy around producing and distributing cannabis products, it is imperative to examine who stands to benefit from its decriminalization, with a focus on land and capital access across racial, class, and gender lines.

16 Smith 2023.

17 Blanchflower et al. 2003.

18 Heir's property is "family-owned land that is jointly owned by descendants of a deceased person," and this practice is found more commonly in places with high poverty and non-white populations. It has been attributed to the generational loss of property, especially in Black communities, by complicating the generational accumulation of wealth through property ownership. See more in Gibson 2022.

19 Glennie 2020.

20 Eizebnberg 2011.

21 Irazábal and Punja 2009.

22 Aptekar 2015.

23 Schilling and Logan 2008; Pothukuchi 2018.

24 The Dudley Neighbors, Inc., Community Land Trust in Boston, managed by Dudley Street Neighborhood Initiative, is considered one of the most successful models of CLT in the US. It has successfully empowered and stabilized a formerly struggling neighborhood, as evidenced by its survival of the 2008 financial crisis without a foreclosure. The CLT includes urban farms and gardens, in addition to over 200 affordable housing units, playgrounds, and commercial spaces. But with its scale and power of eminent domain, the Dudley Street CLT's model may not be replicable in other cities.

25 Rosenberg and Yuen 2012.

26 Foster and Iaione 2022.

27 Ibid.

28 Cahen et al. 2019.

29 Pagano 2013; Douglas 2014; Volont 2019.

30 Chang n.d.

31 Douglas 2014.

METHODOLOGICAL APPENDIX

1 Hermanowicz 2013.

2 Kusenbach 2003.

3 Glaser and Staruss 1967.

4 Barron Ausbrooks et al. 2008; Van Brown 2020.

REFERENCES

Adams, Jane, and D. Gordon. 2009. "This Land Ain't My Land: The Eviction of Share-croppers by the Farm Security Administration." *Agricultural History* 83 (3): 323–51.

Adams, Matthew, and Jayne Raisborough. 2010. "Making a Difference: Ethical Consumption and the Everyday." *The British Journal of Sociology* 61 (2): 256–74.

Adams, Thomas Jessen, and Matt Sakakeeny. 2019. *Remaking New Orleans: Beyond Exceptionalism and Authenticity.* Durham, NC: Duke University Press.

Adams, Vincanne. 2012. "The Other Road to Serfdom: Recovery by the Market and the Affect Economy in New Orleans." *Public Culture* 24 (1 [66]): 185–216.

———. 2013. *Markets of Sorrow, Labors of Faith: New Orleans in the Wake of Katrina.* Durham, NC: Duke University Press.

Airriess, Christopher A., and David L. Clawson. 1994. "Vietnamese Market Gardens in New Orleans." *Geographical Review* 84 (1): 16–31.

Akers, Joshua M. 2012. "Separate and Unequal: The Consumption of Public Education in Post-Katrina New Orleans." *International Journal of Urban and Regional Research* 36 (1): 29–48.

Alexander, Kevin. 2019. *Burn the Ice: The American Culinary Revolution and Its End.* New York: Penguin Press.

Alkon, Alison Hope, Yuki Kato, and Joshua Sbicca. 2020. *A Recipe for Gentrification: Food, Power, and Resistance in the City.* New York: New York University Press.

Allen, Patricia. 2008. "Mining for Justice in the Food System: Perceptions, Practices and Possibilities." *Agriculture and Human Values* 25 (2): 157–61.

Alvarez, R. Michael, Thad E. Hall, and Morgan H. Llewellyn. 2008. "Are Americans Confident Their Ballots Are Counted?" *The Journal of Politics* 70 (3): 754–66.

Aptekar, Sofya. 2015. "Visions of Public Space: Reproducing and Resisting Social Hierarchies in a Community Garden." *Sociological Forum* 30 (1): 209–27.

Arena, John. 2012. *Driven from New Orleans: How Nonprofits Betray Public Housing and Promote Privatization.* Minneapolis: University of Minnesota Press.

Bankston, Carl L. 1998. "Versailles Village: The History and Structure of a Vietnamese Community in New Orleans." *Free Inquiry in Creative Sociology* 26 (1): 79–89.

Barron Ausbrooks, Carrie Y., Edith J. Barrett, and Maria Martinez-Cosio. 2009. "Ethical Issues in Disaster Research: Lessons from Hurricane Katrina." *Population Research and Policy Review* 28 (1): 93–106.

Bassett, T. J. 1981. "Reaping on the Margins: A Century of Community Gardening in America." *Landscape* 25 (2): 1–8.

Benford, Robert D., and David A. Snow. 2000. "Framing Processes and Social Move-ments: An Overview and Assessment." *Annual Review of Sociology* 26 (1): 611–39.

Benke, Kurt, and Bruce Tomkins. 2017. "Future Food-Production Systems: Vertical Farming and Controlled-Environment Agriculture." *Sustainability: Science, Practice and Policy* 13 (1): 13–26.

Bherer, Laurence, Pascale Dufour, and Françoise Montambeault. 2024. "Creating Local 'Citizen's Governance Spaces' in Austerity Contexts: Food Recuperation and Urban Gardening in Montréal (Canada) as Ways to Pragmatically Invent Alternatives." *Urban Affairs Review* DOI: 10780874231224359.

Blanchflower, David G, Phillip B Levine, and David J Zimmerman. 2003. "Discrimina-tion in the Small-Business Credit Market." *The Review of Economics and Statistics* 85 (4): 930–43.

Boggs, Carl. 1977. "Marxism, Prefigurative Communism, and the Problem of Workers' Control." *Radical America* 11 (6): 99–122.

Born, Branden, and Mark Purcell. 2006. "Avoiding the Local Trap: Scale and Food Systems in Planning Research." *Journal of Planning Education and Research* 26 (2): 195–207.

Brand, Anna Livia, and Vern Baxter. 2020. "Post-Disaster Development Dilemmas: Advancing Landscapes of Social Justice in a Neoliberal Post-Disaster Landscape." In *Louisiana's Response to Extreme Weather: A Coastal State's Adaptation Challenges and Successes*, edited by Shirley Laska, 217–40. New York: Springer Nature.

Broad, Garrett. 2016. *More than Just Food: Food Justice and Community Change*. Berke-ley: University of California Press.

Brones, Anna. 2018. "Karen Washington: It's Not a Food Desert, It's Food Apartheid." *Guernica*, May 7, 2018. https://www.guernicamag.com/karen-washington-its-not-a-food-desert-its-food-apartheid/.

Broom, Pamela, and Yuki Kato. 2020. "The Holy Trinity to Microgreens: Gentrification Redefining the Local Foodways." In *A Recipe for Gentrification: Food, Power, and Resis-tance in the City*, edited by Alison Hope Alkon, Yuki Kato, and Joshua Sbicca, 111–31.

Broom, Pamela, Yuki Kato, and Shawn "Pepper" Roussel. 2024. "Rewriting the Erased History of Blacks in New Orleans Urban Gardening and Farming." *Humanity & Society*. DOI: 10.1177/01605976241266239

Brown, M Kay. 2018. "The Power of Leaving: Black Agency and the Great Migration in Louisiana, 1890–1939." Undergraduate Honors Thesis, New Orleans: University of New Orleans.

Brown-Saracino, Japonica. 2016. "An Agenda for the next Decade of Gentrification Scholarship." *City & Community* 15 (3): 220–25.

Brunsma, David L., David Overfelt, and J. Steven Picou. 2007. *The Sociology of Katrina: Perspectives on a Modern Catastrophe*. Lanhamm, MD: Rowman & Littlefield Publishers.

Bullard, Robert D., and Beverly Wright. 2009. *Race, Place, and Environmental Justice After Hurricane Katrina: Struggles to Reclaim, Rebuild, and Revitalize New Orleans and the Gulf Coast*. Boulder, CO: Westview Press.

Bunnage, Leslie A. 2014. "Social Movement Engagement over the Long Haul: Understanding Activist Retention." *Sociology Compass* 8 (4): 433–45. https://doi.org/10.1111/soc4.12141.

Cahen, Claire, Jakob Schneider, and Susan Saegert. 2019. "Victories from Insurgency: Re-Negotiating Housing, Community Control, and Citizenship at the Margins." *Antipode* 51 (5): 1416–35.

Campanella, Richard. 2009. "'Bring Your Own Chairs': Civic Engagement in Postdiluvial New Orleans." In *Civic Engagement in the Wake of Katrina,* edited by Amy Koritz and George Sanchez, 23–42. The Hague: Oapen.

———. 2013. "Gentrification and Its Discontents: Notes from New Orleans." *New Geography* 3.

———. 2020. "Now Forgotten, 'Truck Farms' Once Dotted New Orleans, but Were Overtaken by Urban Growth." *The Times-Picayune*, October 1, 2020.

Campbell, Emahunn Raheem Ali. 2011. "A Critique of the Occupy Movement from a Black Occupier." *The Black Scholar* 41 (4): 42–51.

Campbell, Lindsay K. 2017. *City of Forests, City of Farms: Sustainability Planning for New York City's Nature.* Ithaca: Cornell University Press.

Cannon, C. W. 2014. "A Kale of Two Cities: The Magical New Orleans and the Americanist Version." *The Lens* (blog). March 14, 2014. https://thelensnola.org/.

Ceaser, Donovon. 2012. "Our School at Blair Grocery: A Case Study in Promoting Environmental Action Through Critical Environmental Education." *The Journal of Environmental Education* 43 (4): 209–26.

"Celebrate Our History, Invest in Our Future: Reinvigorating Tourism in New Orleans." 2010. New Orleans, LA: New Orleans Strategic Hospitality Task Force. https://neworleanscitybusiness.com/wp-files/hospitality-plan.pdf

Chang, Candy. n.d. "I Wish This Was." n.d. https://candychang.com/work/i-wish-this-was/.

Checker, Melissa. 2020. *The Sustainability Myth: Environmental Gentrification and the Politics of Justice.* New York: New York University Press.

Chen, Cher Weixia, and Paul C. Gorski. 2015. "Burnout in Social Justice and Human Rights Activists: Symptoms, Causes and Implications." *Journal of Human Rights Practice* 7 (3): 366–90.

Cherrie, Lolita V. 2014. "The Building of the Lafitte Housing Project—1941." August 20, 2014. http://www.creolegen.org/2014/08/20/the-building-of-the-lafitte-housing-project-1941.

Cho, Albert Hyunbae. 2006. "Politics, Values and Social Entrepreneurship: A Critical Appraisal." In *Social Entrepreneurship*, edited by Johanna Mair, Jeffrey Robinson, and Kai Hockerts, 34–56. New York: Palgrave MacMillan.

Clark, Jack. 2010. "ACORN's Demise." *Dissent Magazine* (blog). October 27, 2010.

Corrigall-Brown, Catherine. 2011. *Patterns of Protest: Trajectories of Participation in Social Movements.* Stanford: Stanford University Press.

Dacin, M. Tina, Peter A. Dacin, and Paul Tracey. 2011. "Social Entrepreneurship: A Critique and Future Directions." *Organization Science* 22 (5): 1203–13. https://doi.org/10.1287/orsc.1100.0620.

D'Andrade, Roy G. 1992. "Schemas and Motivation." In *Human Motives and Cultural Models*, edited by Roy G. D'Andrade and Claudia Strauss, 23–44. Cambridge, UK: Cambridge University Press.

Dees, J. Gregory. 2018. "The Meaning of Social Entrepreneurship 1, 2." In *Case Studies in Social Entrepreneurship and Sustainability*, edited by Jost Hamschmidt and Michael Pirson, 1st ed., 22–30. London: Routledge.

DeLind, Laura B. 2011. "Are Local Food and the Local Food Movement Taking Us Where We Want to Go? Or Are We Hitching Our Wagons to the Wrong Stars?" *Agriculture and Human Values* 28 (2): 273–83.

Densmore, Frances. 1944. "A Search for Songs Among the Chitimacha Indians in Louisiana." Anthropological Papers No. 19. Bulletin 133. Bureau of American Ethnology, Smithsonian Institution. https://repository.si.edu/bitstream/handle/10088/34582/bae_bulletin_133_1943_No19.pdf

Despommier, Dickson. 2020. *The Vertical Farm: Feeding the World in the Twenty-First Century*. New York: Macmillan.

Dooling, Sarah. 2009. "Ecological Gentrification: A Research Agenda Exploring Justice in the City." *International Journal of Urban and Regional Research* 33 (3): 621–39.

Douglas, Gordon C. C. 2014. "Do-It-Yourself Urban Design: The Social Practice of Informal 'Improvement' Through Unauthorized Alteration." *City & Community* 13 (1): 5–25.

Douglas, Lake. 2011. *Public Spaces, Private Gardens: A History of Designed Landscapes in New Orleans*. Baton Rouge: Louisiana State University Press.

Draus, Paul J. 2009. "Substance Abuse and Slow-Motion Disasters: The Case of Detroit." *The Sociological Quarterly* 50 (2): 360–82.

DuPuis, Melanie E., and David Goodman. 2005. "Should We Go 'Home' to Eat?: Toward a Reflexive Politics of Localism." *Journal of Rural Studies* 21 (3): 359–71.

Dyson, Michael Eric. 2006. *Come Hell or High Water*. New York: Basic Civitas Books.

Eagleton, Terry. 2015. *Hope without Optimism*. Charlottesville: University of Virginia Press.

Egendorf, Sara Perl, Peter Groffman, Gerry Moore, and Zhongqi Cheng. 2020. "The Limits of Lead (Pb) Phytoextraction and Possibilities of Phytostabilization in Contaminated Soil: A Critical Review." *International Journal of Phytoremediation* 22 (9): 916–30.

Ehrenfeucht, Renia, and Marla Nelson. 2013. "Young Professionals as Ambivalent Change Agents in New Orleans after the 2005 Hurricanes." *Urban Studies* 50 (4): 825–41.

Eizenberg, Efrat. 2012. "The Changing Meaning of Community Space: Two Models of NGO Management of Community Gardens in New York City." *International Journal of Urban and Regional Research* 36 (1): 106–20.

Engler, Nicholas, and Moncef Krarti. 2021. "Review of Energy Efficiency in Controlled Environment Agriculture." *Renewable and Sustainable Energy Reviews* 141 (May): 110786.

Erikson, Kai T. 1976. *Everything in Its Path: Destruction of Community in the Buffalo Creek Flood*. New York: Simon and Schuster.

Finley, Laura, and Luigi Esposito. 2012. "Neoliberalism and the Non-Profit Industrial Complex: The Limits of a Market Approach to Service Delivery." *Peace Studies Journal* 5 (November): 4–26.

Finn, Donovan, and Gordon Douglas. 2019. "DIY Urbanism." in *A Research Agenda for New Urbanism* edited by Emily Talen. Northampton, MA: Edward Elgar Publishing. 20–34.

Firth, Jeanne K. 2023. *Feeding New Orleans: Celebrity Chefs and Reimagining Food Justice*. Chapel Hill: University of North Carolina Press.

Flaherty, Jordan. 2010. *Floodlines: Community and Resistance from Katrina to the Jena Six*. Chicago: Haymarket Books.

———. 2016. *No More Heroes: Grassroots Challenges to the Savior Mentality*. Chico, CA: AK Press.

Florida, Richard. 2002. *The Rise of the Creative Class*. New York: Basic Books.

Foster, Sheila R., and Christian Iaione. 2022. *Co-Cities: Innovative Transitions toward Just and Self-Sustaining Communities*. Cambridge, MA: The MIT Press.

Freudenburg, William R., Robert B. Gramling, Shirley Laska, and Kai Erikson. 2009. *Catastrophe in the Making: The Engineering of Katrina and the Disasters of Tomorrow*. Washington, DC: Island Press.

Fussell, Elizabeth. 2007. "Constructing New Orleans, Constructing Race: A Population History of New Orleans." *Journal of American History* 94 (3): 846–55. https://doi.org /10.2307/25095147.

———. 2015. "The Long-Term Recovery of New Orleans' Population after Hurricane Katrina." *American Behavioral Scientist* 59 (10): 1231-1245.

Fussell, Elizabeth, Narayan Sastry, and Mark VanLandingham. 2010. "Race, Socioeconomic Status, and Return Migration to New Orleans after Hurricane Katrina." *Population and Environment* 31 (1): 20–42.

Gessler, Anne. 2020. *Cooperatives in New Orleans: Collective Action and Urban Development*. Jackson, MS: University Press of Mississippi.

Gibson, Brenda D. 2022. "The Heirs' Property Problem: Racial Caste Origins & Systemic Effects in the Black Community." CUNY Law Review 26: 172.

Gill, Duane A. 2007. "Secondary Trauma or Secondary Disaster? Insights from Hurricane Katrina." *Sociological Spectrum* 27 (6): 613–32.

Glaser, Barney G., and Anselm L. Strauss. 1967. *The Discovery of Grounded Theory: Strategies for Qualitative Research*. Piscataway, NJ: Transaction Publishers.

Glennie, Charlotte. 2020. "Cultivating Place: Urban Development and the Institutionalization of Seattle's P-Patch Community Gardens." *City & Community* 19 (3): 726–46.

Gorski, Paul C. 2019a. "Racial Battle Fatigue and Activist Burnout in Racial Justice Activists of Color at Predominately White Colleges and Universities." *Race Ethnicity and Education* 22 (1): 1–20.

———. 2019b. "Fighting Racism, Battling Burnout: Causes of Activist Burnout in US Racial Justice Activists." *Ethnic and Racial Studies* 42 (5): 667–87. https://doi.org/10 .1080/01419870.2018.1439981.

Gorski, Paul, Stacy Lopresti-Goodman, and Dallas Rising. 2019. "'Nobody's Paying Me to Cry': The Causes of Activist Burnout in United States Animal Rights Activists." *Social Movement Studies* 18 (3): 364–80.

Gotham, Kevin F. 2007. *Authentic New Orleans: Tourism, Culture, and Race in the Big Easy*. New York: New York University Press.

———. 2012. "Disaster, Inc.: Privatization and Post-Katrina Rebuilding in New Orleans." *Perspectives on Politics* 10 (3): 633–46.

———. 2014. "Racialization and Rescaling: Post-Katrina Rebuilding and the Louisiana Road Home Program." *International Journal of Urban and Regional Research* 38 (3): 773–90.

———. 2015. "Limitations, Legacies, and Lessons: Post-Katrina Rebuilding in Retrospect and Prospect." *American Behavioral Scientist* 59 (10): 1314–26.

———. 2017. "Touristic Disaster: Spectacle and Recovery in Post-Katrina New Orleans." *Geoforum* 86: 127–35.

Gotham, Kevin Fox, and Miriam Greenberg. 2014. *Crisis Cities: Disaster and Redevelopment in New York and New Orleans*. New York: Oxford University Press.

Gould, Kenneth A., and Tammy L. Lewis. 2016. *Green Gentrification: Urban Sustainability and the Struggle for Environmental Justice*. New York: Routledge.

Granovetter, Mark S. 1973. "The Strength of Weak Ties." *American Journal of Sociology* 78 (6): 1360–80.

Greenberg, Jason. 2019. "Inequality and Crowdfunding." In *Handbook of Research on Crowdfunding*, edited by Hans Landström, Annaleena Parhankangas, and Colin Mason, 303–22. Cheltenham, UK and Northampton, MA: Edward Elgar Publishing.

Guthman, Julie. 2008. "Bringing Good Food to Others: Investigating the Subjects of Alternative Food Practice." *Cultural Geographies* 15 (4): 431–47.

Haenfler, Ross, Brett Johnson, and Ellis Jones. 2012. "Lifestyle Movements: Exploring the Intersection of Lifestyle and Social Movements." *Social Movement Studies* 11 (1): 1–20.

Hartman, Chester W., and Gregory D. Squires. 2006. *There Is No Such Thing as a Natural Disaster: Race, Class, and Hurricane Katrina*. New York: Routledge.

Harvey, Daina Cheyenne, Yuki Kato, and Catarina Passidomo. 2016. "Rebuilding Others' Communities: A Critical Analysis of Race and Nativism in Non-Profits in the Aftermath of Hurricane Katrina." *Local Environment* 21 (8): 1029–46.

Harvey, David. 2007. *A Brief History of Neoliberalism*. New York: Oxford University Press.

Hassberg, Analena Hope. 2020. "Citified Sovereignty: Cultivating Autonomy in South Los Angeles." In *A Recipe for Gentrification: Food, Power, and Resistance in the City*, edited by Alison H. Alkon, Yuki Kato, and Joshua Sbicca, 305–24. New York: New York University Press.

Haydu, Jeffrey. 2021. *Upsetting Food: Three Eras of Food Protests in the United States*. Philadelphia: Temple University Press.

Hensby, Alexander. 2021. "Political Non-participation in Elections, Civic Life and Social Movements." *Sociology Compass* 15 (1).

Hermanowicz, Joseph C. 2013. "The Longitudinal Qualitative Interview." *Qualitative Sociology* 36 (2): 189–208.

Hinrichs, C. Clare. 2000. "Embeddedness and Local Food Systems: Notes on Two Types of Direct Agricultural Market." *Journal of Rural Studies* 16 (3): 295–303.

Holm, Eric Joseph van, and Christopher K. Wyczalkowski. 2019. "Gentrification in the Wake of a Hurricane: New Orleans after Katrina." *Urban Studies* 56 (13): 2763–78.

Holt Giménez, Eric, and Annie Shattuck. 2011. "Food Crises, Food Regimes and Food Movements: Rumblings of Reform or Tides of Transformation?" *The Journal of Peasant Studies* 38 (1): 109–44.

Hondagneu-Sotelo, Pierrette. 2014. *Paradise Transplanted: Migration and the Making of California Gardens.* Berkeley: University of California Press.

Horowitz, Andy. 2020. *Katrina: A History, 1915–5015.* Cambridge, MA: Harvard University Press.

Hou, Jeffrey. 2010. *Insurgent Public Space: Guerrilla Urbanism and the Remaking of Contemporary Cities.* New York: Routledge.

INCITE! Women of Color Against Violence. 2017. *The Revolution Will Not Be Funded. The Revolution Will Not Be Funded.* Durham, NC: Duke University Press.

Irazábal, Clara, and Anita Punja. 2009. "Cultivating Just Planning and Legal Institutions: A Critical Assessment of the South Central Farm Struggle in Los Angeles." *Journal of Urban Affairs* 31 (1): 1–23.

Irvin, Cate. 2017. "Constructing Hybridized Authenticities in the Gourmet Food Truck Scene." *Symbolic Interaction* 40 (1): 43–62.

Jacobs, Jane. 1961. *The Death and Life of Great American Cities.* New York: Vintage Books.

Jeffrey, Craig, and Jane Dyson. 2021. "Geographies of the Future: Prefigurative Politics." *Progress in Human Geography* 45 (4): 641–58.

Jensen, Lynne. 1986. "Agriculture Sprouts in Inner City." *New Orleans Times-Picayune,* February 6, 1986, sec. East New Orleans.

Johnson, Cedric, ed. 2011. *The Neoliberal Deluge: Hurricane Katrina, Late Capitalism, and the Remaking of New Orleans.* Minneapolis: University of Minnesota Press.

———. 2015. "Gentrifying New Orleans: Thoughts on Race and the Movement of Capital." *Souls* 17 (3–4): 175–200.

Johnston, Josée, and Shyon Baumann. 2010. *Foodies: Democracy and Distinction in the Gourmet Foodscape.* New York: Routledge.

Jordan, Grant, and William Maloney. 2006. "'Letting George Do It': Does Olson Explain Low Levels of Participation?" *Journal of Elections, Public Opinion and Parties* 16 (2): 115–39.

Kato, Yuki. 2013. "Not Just the Price of Food: Challenges of an Urban Agriculture Organization in Engaging Local Residents." *Sociological Inquiry* 83 (3): 369–91.

———. 2020. "Gardening in Times of Urban Transitions: Emergence of Entrepreneurial Cultivation in Post-Katrina New Orleans." *City & Community* 19 (4): 987–1010.

Kato, Yuki, Scarlett Andrews, and Cate Irvin. 2018. "Availability and Accessibility of Vacant Lots for Urban Cultivation in Post-Katrina New Orleans." *Urban Affairs Review* 54 (2): 322–62.

Keller, Judith. 2022. "Brad Pitt's Apparently Defunct Foundation Reached a $20.5 Million Settlement with Hurricane Katrina Survivors over Its Green Housing Debacle." *The Conversation* (blog). August 23, 2022. http://theconversation.com/.

Klein, Naomi. 2008. *The Shock Doctrine: The Rise of Disaster Capitalism*. New York: Picador.

Kluckow, Richard. 2014. "The Impact of Heir Property on Post-Katrina Housing Recovery in New Orleans" PhD dissertation, Colorado State University. ProQuest (1573070).

Kotkin, Joel. 2014. "Sustaining Prosperity: A Long Term Vision for the New Orleans Region." Greater New Orleans, Inc. http://gnoinc.org/uploads/Sustaining_Prosperity_Amended_2014_02_16.pdf. Accessed August 15, 2015.

Kusenbach, Margarethe. 2003. "Street Phenomenology: The Go-Along as Ethnographic Research Tool." *Ethnography* 4 (3): 455–85.

LaFrombois, Megan Heim. 2017. "Blind Spots and Pop-up Spots: A Feminist Exploration into the Discourses of Do-It-Yourself (DIY) Urbanism." *Urban Studies* 54 (2): 421–36.

Landphair, Juliette. 2007. "'The Forgotten People of New Orleans': Community, Vulnerability, and the Lower Ninth Ward." *The Journal of American History* 94 (3): 837–45.

Langenhennig, Susan. 2015. "Seeds and the City: Urban Lots Are Sprouting Farms across New Orleans." *nola.com*. June 17, 2015.

Lawson, Laura. 2005. *City Bountiful: A Century of Community Gardening in America*. Berkeley: University of California Press.

Lees, Loretta, and Martin Phillips. 2018. *Handbook of Gentrification Studies*. Cheltenham, UK and Northampton, MA: Edward Elgar Publishing.

Leong, Karen J., Christopher A. Airriess, Wei Li, Angela Chia-Chen Chen, and Verna M. Keith. 2007. "Resilient History and the Rebuilding of a Community: The Vietnamese American Community in New Orleans East." *The Journal of American History* 94 (3): 770–79.

Library of Congress. n.d. "Community Agricultural Programs & Urban Food Hubs." *Research Guides* (blog). n.d. https://guides.loc.gov/community-agricultural-programs-urban-food-hubs/plant-production/funding. Accessed August 20, 2015.

Lindemann, Justine. 2022. "'A Little Portion of Our 40 Acres': A Black Agrarian Imaginary in the City." *Environment and Planning E: Nature and Space* DOI: 25148486221129410.

Litwin, Sharon. 1981. "They're Planting Their Futures." *New Orleans Times-Picayune*, May 12, 1981.

Logan, John, and Harvey Molotch. 1987. *Urban Fortunes: The Political Economy of Place*. Berkeley: University of California Press.

Lueck, Michelle A. M. 2007. "Hope for a Cause as Cause for Hope: The Need for Hope in Environmental Sociology." *The American Sociologist* 38 (3): 250–61.

Luft, Rachel E. 2008. "Looking for Common Ground: Relief Work in Post-Katrina New Orleans as an American Parable of Race and Gender Violence." *NWSA Journal* 20 (3): 5–31.

———. 2009. "Beyond Disaster Exceptionalism: Social Movement Developments in New Orleans after Hurricane Katrina." *American Quarterly* 61 (3): 499–527.

Margavio, A.V., and Jerome Salomone. 2014. *Bread and Respect: The Italians of Louisiana*. New Orleans: Pelican Publishing.

Mason, Kelvin. 2014. "Becoming Citizen Green: Prefigurative Politics, Autonomous Geographies, and Hoping against Hope." *Environmental Politics* 23 (1): 140–58.

Massey, Douglas S., and Nancy A. Denton. 1993. *American Apartheid: Segregation and the Making of the Underclass*. Cambridge, MA: Harvard University Press.

"Master Plan (Public Comments) Attachment A, Volume 2, Chapter 12: Land Use Plan." 2011. City of New Orleans. Accessed April 12, 2014.

Mayer, Vicki, and Tanya Goldman. 2010. "Hollywood Handouts: Tax Credits in the Age of Economic Crisis." *Jump Cut* 52 (Summer).

Mayes, Christopher. 2014. "An Agrarian Imaginary in Urban Life: Cultivating Virtues and Vices Through a Conflicted History." *Journal of Agricultural and Environmental Ethics* 27 (2): 265–86.

McAdam, Doug. 1986. "Recruitment to High-Risk Activism: The Case of Freedom Summer." *American Journal of Sociology* 92 (1): 64–90.

McClintock, Nathan. 2010. "Why Farm the City? Theorizing Urban Agriculture through a Lens of Metabolic Rift." *Cambridge Journal of Regions, Economy and Society* 3 (2): 191–207.

———. 2014. "Radical, Reformist, and Garden-Variety Neoliberal: Coming to Terms with Urban Agriculture's Contradictions." *Local Environment* 19 (2): 147–71.

McGeer, Victoria. 2004. "The Art of Good Hope." *The ANNALS of the American Academy of Political and Social Science* 592 (1): 100–127.

McKeever, Brice, and Thomas H. Pollak. 2015. "Ten Years after Katrina, New Orleans Nonprofits Are Still Growing." Urban Wire. Urban Institute. https://www.urban.org/urban-wire/ten-years-after-katrina-new-orleans-nonprofits-are-still-growing.

McMillan Cottom, Tressie. 2020. "Virtues & Vocations—Race and Purpose in Higher Education with Tressie McMillan Cottom." Kenan Institute for Ethics, Duke University (Lecture via Zoom), August 25.

———. 2021. "Keynote Speaker." Presented at the Association of College & Research Libraries, Virtual Meeting, April 13. https://www.libraryjournal.com/story/Keynoter-Tressie-McMillan-Cottom-Talks-Human-Centered-Data-Rights-and-Pragmatic-Hope-ACRL-2021.

Midgley, Ian, dir. 2015. *Reversing the Mississippi*. https://www.documentary.org/project/reversing-mississippi.

Miller, R.G. 1918. "Wheels of Progress." New Orleans States. 99–57-L. The Historic New Orleans Collection's World War I Scrapbook. https://catalog.hnoc.org/web/arena/search#/entity/thnoc-archive/99-57-L/world-war-i-scrapbook.

Minkoff-Zern, Laura-Anne. 2014. "Challenging the Agrarian Imaginary: Farmworker-Led Food Movements and the Potential for Farm Labor Justice." *Human Geography* 7 (1): 85–101.

Monteiro, Renato, José C. Ferreira, and Paula Antunes. 2020. "Green Infrastructure Planning Principles: An Integrated Literature Review." *Land* 9 (12): 525.

Montgomery, Alesia. 2020. *Greening the Black Urban Regime: The Culture and Commerce of Sustainability in Detroit.* Detroit: Wayne State University Press.

Moskowitz, Peter. 2015. "New Orleans' Lower Ninth Ward Targeted for Gentrification: 'It's Going to Feel like It Belongs to the Rich.'" *Guardian*, January 23, 2015, sec. US news.

Mwendo, Jenga. 2012. "Jungleland? New Orleans Community Activist Rejects NY Times Depiction of 9th Ward." *EBONY*, May 9, 2012. https://www.ebony.com/.

Myers, Justin Sean. 2022. *Growing Gardens, Building Power: Food Justice and Urban Agriculture in Brooklyn.* New Brunswick, NJ: Rutgers University Press.

National Young Farmers Coalition. 2018. "A Farm Bill for the Future." *National Young Farmers Coalition* (blog). December 12, 2018. https://www.youngfarmers.org/2018/12/farmbillforthefuture/.

Nettle, Claire. 2014. *Community Gardening as Social Action.* Burlington, VT: Ashgate Publishing Company.

New Orleans Redevelopment Authority. n.d. "Growing Green: Summary Information." New Orleans Redevelopment Authority.

Ocejo, Richard E. 2017. *Masters of Craft: Old Jobs in the New Urban Economy.* Princeton, NJ: Princeton University Press.

Oldenburg, Ray. 1989. *The Great Good Place: Cafés, Coffee Shops, Community Centers, Beauty Parlors, General Stores, Bars, Hangouts, and How They Get You through the Day.* New York: Paragon House.

Olson Jr, Mancur. 1971. *The Logic of Collective Action: Public Goods and the Theory of Groups, with a New Preface and Appendix.* Vol. 124. Harvard University Press.

Pack, Charles Lathrop. 1919. *The War Garden Victorious.* JP Lippincott Company.

Pagano, Celeste. 2013. "DIY Urbanism: Property and Process in Grassroots City Building." *Marquette Law Review* 97 (2): 335–90.

Palmer, Vernon Valentine. 2006. "The Customs of Slavery: The War Without Arms." *Global Jurist Frontiers* 6 (1).

Passidomo, Catarina. 2016. "Community Gardening and Governance over Urban Nature in New Orleans's Lower Ninth Ward." *Urban Forestry & Urban Greening*, Special Section: Power in urban social-ecological systems: Processes and practices of governance and marginalization, 19 (September): 271–77.

Patel, Megha M., Leia Y. Saltzman, Regardt J. Ferreira, and Amy E. Lesen. 2018. "Resilience: Examining the Impacts of the Deepwater Horizon Oil Spill on the Gulf Coast Vietnamese American Community." *Social Sciences* 7 (10): 203.

Peck, Jamie. 2005. "Struggling with the Creative Class." *International Journal of Urban and Regional Research* 29 (4): 740–70.

Penniman, Leah. 2018. *Farming While Black: Soul Fire Farm's Practical Guide to Liberation on the Land.* White River Junction, VT: Chelsea Green Publishing Company.

Peredo, Ana María, and Murdith McLean. 2006. "Social Entrepreneurship: A Critical Review of the Concept." *Journal of World Business* 41 (1): 56–65.

Peters, Xander. 2020. "What Happened to New Orleans' Black Truck Farming Culture?" *Scalawag* (blog). April 27, 2020. https://scalawagmagazine.org/2020/04/new-orleans-black-truck-agriculture/.

Pew Research Center. 2023. "Public Trust in Government: 1958–2023." September 19, 2023. https://www.pewresearch.org/politics/2023/09/19/public-trust-in-government-1958-2023/.

Picou, J. Steven. 2009. "Katrina as a Natech Disaster: Toxic Contamination and Long-Term Risks for Residents of New Orleans." *Journal of Applied Social Science* 3 (2): 39–55.

Plyer, Allison, and Elaine Ortiz. 2012. "Benchmarks for Blight: How Much Blight Does New Orleans Have?" Report. Greater New Orleans Community Data Center. http://www.gnocdc.org/BenchmarksForBlight/index.html. Accessed May 4, 2013.

Polletta, Francesca, and James M. Jasper. 2001. "Collective Identity and Social Movements." *Annual Review of Sociology* 27 (1): 283–305.

Pothukuchi, Kameshwari. 2017. "'To Allow Farming Is to Give up on the City': Political Anxieties Related to the Disposition of Vacant Land for Urban Agriculture in Detroit." *Journal of Urban Affairs* 39 (8): 1169–89.

———. 2018. "Vacant Land Disposition for Agriculture in Cleveland, Ohio: Is Community Development a Mixed Blessing?" *Journal of Urban Affairs* 40 (5): 657–78.

Pottharst, Kris. 1995. "Urban Dwellers and Vacant Lots." *Parks & Recreation* 30 (9): 94–102.

Pottinger, Laura. 2017. "Planting the Seeds of a Quiet Activism." *Area* 49 (2): 215–22.

Propeller. n.d. "Lots of Progress PitchNOLA Competition." Propeller. Accessed March 21, 2020. http://gopropeller.org/events/126/.

Putnam, Robert D. 2000. *Bowling Alone: The Collapse and Revival of American Community*. New York: Simon and Schuster.

Rabito, Felicia A., Shahed Iqbal, Sara Perry, Whitney Arroyave, and Janet C. Rice. 2012. "Environmental Lead after Hurricane Katrina: Implications for Future Populations." *Environmental Health Perspectives* 120 (2): 180–84.

Reeves, Sally Kittredge. 2001. *Jacques-Felix Lelievre's New Louisiana Gardener*. Baton Rouge, LA: Louisiana State University Press.

Reynolds, Kristin, and Nevin Cohen. 2016. *Beyond the Kale*. Athens, GA: University of Georgia Press.

Rich, Nathaniel. 2012. "Jungleland." *New York Times*, March 21, 2012, sec. Magazine. https://www.nytimes.com/2012/03/25/magazine/the-lower-ninth-ward-new-orleans.html.

Rosan, Christina D., and Hamil Pearsall. 2017. *Growing a Sustainable City?: The Question of Urban Agriculture*. Toronto: University of Toronto Press.

Rose, Donald J., Nicholas Bodor, Chris M. Swalm, Janet C. Rice, Thomas A. Farley, and Paul I. Hutchinson. 2009. "Deserts in New Orleans?: Illustrations of Urban Food Access and Implications for Policy." Report. University of Michigan National Poverty Center/USDA Economic Research Service Research. https://npc.umich.edu/news/events/food-access/rose_et_al.pdf. Accessed February 28, 2014.

Rosenberg, Greg, and Jeffrey Yuen. 2012. "Beyond Housing: Urban Agriculture and Commercial Development by Community Land Trusts." Lincoln Institute of Land Policy. https://www.lincolninst.edu/app/uploads/2024/04/2227_1559_rosenberg _wp13gr1.pdf

Rosenberg, Nathan A., and Bryce Wilson Stucki. 2017. "The Butz Stops Here: Why the Food Movement Needs to Rethink Agricultural History." *Journal of Food Law and Policy* 13: 12.

Rothstein, Richard. 2017. *The Color of Law: A Forgotten History of How Our Government Segregated America*. New York: W.W. Norton.

Saldivar-Tanaka, Laura, and Marianne E. Kransy. 2004. "Culturing Community Development, Neighborhood Open Space, and Civic Agriculture: The Case of Latino Community Gardens in New York City." *Agriculture and Human Values* 21 (4): 399–412.

Sbicca, Joshua. 2018. *Food Justice Now!: Deepening the Roots of Social Struggle*. Minneapolis: University of Minnesota Press.

———. 2019. "Urban Agriculture, Revalorization, and Green Gentrification in Denver, Colorado." *The Politics of Land*, 26:149–70.

Schilling, Joseph, and Jonathan Logan. 2008. "Greening The Rust Belt: A Green Infrastructure Model For Right Sizing America's Shrinking Cities." *Journal of the American Planning Association* 74 (4): 451–66.

Schoolman, Ethan D. 2019. "Doing Right and Feeling Good: Ethical Food and the Shopping Experience." *Sociological Perspectives* 62 (5): 668–90.

Schor, Juliet B. 2020. *After the Gig: How the Sharing Economy Got Hijacked and How to Win It Back*. Oakland, California: University of California Press.

Scott, John H., and Cleo Scott Brown. 2008. *Witness to the Truth—HFS Books*. Columbia: University of South Carolina Press.

Shostak, Sara. 2021. Back to the Roots: Memory, Inequality, and Urban Agriculture. New Brunswick, NJ: Rutgers University Press.

———. 2022. "'How Do We Measure Justice?': Missions and Metrics in Urban Agriculture." Agriculture and Human Values 39 (3): 953–64. https://doi.org/10.1007/s10460 -022-10296-4.

Silverman, Robert Mark, Kelly L. Patterson, and Samantha Shavon Williams. 2023. "Don't Fear the Reefer? The Social Equity and Community Planning Implications of New York'sRecreational Cannabis Law on Underserved Communities." *Journal of Policy Practice and Research* 4 (January): 150–67.

Smith II, Bobby J. 2023. *Food Power Politics: The Food Story of the Mississippi Civil Rights Movement*. Chapel Hill: University of North Carolina Press.

Smith, Clint. 2021. *How the Word Is Passed: A Reckoning with the History of Slavery Across America*. New York: Little, Brown and Company.

Smith, Neil. 1987. "Gentrification and the Rent Gap." *Annals of the Association of American Geographers* 77 (3): 462–65.

Snow, David A, and Robert D. Benford. 1988. "Ideology, Frame Resonance, and Participant Mobilization." *International Social Movement Research* 1 (1): 197–217.

Snow, David A, Rens Vliegenthart, and Pauline Ketelaars. 2019. "The Framing Perspective on Social Movements: Its Conceptual Roots and Architecture." In *The Wiley Blackwell Companion to Social Movements. Oxford: Wiley Blackwell,* edited by David A. Snow, Sarah A. Soule, Hanspeter Kriesi, and Holly J. McCammon, 392–410. Hoboken, NJ: Wiley Blackwell.

Swain, Dan. 2019. "Not Not but Not yet: Present and Future in Prefigurative Politics." *Political Studies* 67 (1): 47–62.

Swidler, Ann. 2003. *Talk of Love: How Culture Matters.* Chicago: University of Chicago Press.

Tang, Eric. 2011. "A Gulf Unites Us: The Vietnamese Americans of Black New Orleans East." *American Quarterly* 63 (1): 117–49.

Taylor, Verta. 1989. "Social Movement Continuity: The Women's Movement in Abeyance." *American Sociological Review* 54 (5): 761–75.

Taylor, Verta, and Leila Rupp. 1987. *Survival in the Doldrums: The American Women's Rights Movement, 1945 to the 1960s.* New York: Oxford University Press.

The Eat Local New Orleans. 2015. "The 5th Annual Eat Local Challenge." http://nolalocavore.org. Accessed June 11, 2015.

"The Lower Ninth Ward Urban Farming Coalition." n.d. http://www.lowernineurbanfarming.org/. Accessed April 10, 2009.

Thomas, Lynnell L. 2009. "'Roots Run Deep Here': The Construction of Black New Orleans in Post-Katrina Tourism Narratives." *American Quarterly* 61 (3): 749–68.

Tierney, Kathleen. 2015. "Resilience and the Neoliberal Project: Discourses, Critiques, Practices—And Katrina." *American Behavioral Scientist* 59 (10): 1327–42.

Tierney, Kathleen, and Anthony Oliver-Smith. 2012. "Social Dimensions of Disaster Recovery." *International Journal of Mass Emergencies and Disasters* 30 (2): 123–46.

Tornaghi, Chiara. 2017. "Urban Agriculture in the Food-Disabling City: (Re)Defining Urban Food Justice, Reimagining a Politics of Empowerment." *Antipode* 49 (3): 781–801. https://doi.org/10.1111/anti.12291.

Tornaghi, Chiara, and Michiel Dehaene. 2020. "The Prefigurative Power of Urban Political Agroecology: Rethinking the Urbanisms of Agroecological Transitions for Food System Transformation." *Agroecology and Sustainable Food Systems* 44 (5): 594–610.

Törnberg, Anton. 2021. "Prefigurative Politics and Social Change: A Typology Drawing on Transition Studies." *Distinktion: Journal of Social Theory* 22 (1): 83–107.

Tremé, season 1, episode 3, "Right Place, Wrong Time," directed by David Simon, aired April 25, 2010, on HBO.

US Census. n.d. "U.S. Census Bureau QuickFacts: New Orleans City, Louisiana." Accessed June 10, 2023. https://www.census.gov/quickfacts/fact/table/neworleanscity louisiana/INC110221.

US Department of Agriculture. 2022. "Urban Growers." US Department of Agriculture. https://www.farmers.gov/your-business/urban-growers. Accessed April 26, 2022.

Van Brown, Bethany L. 2020. "Disaster Research 'Methics': Ethical and Methodological Considerations of Researching Disaster-Affected Populations." *American Behavioral Scientist* 64 (8): 1050–65.

Vickery, Kathryn Koebert. 2014. "Barriers to and Opportunities for Commercial Urban Farming: Case Studies from Austin, Texas and New Orleans, Louisiana." Masters Thesis, The University of Texas at Austin, http://hdl.handle.net/2152/26500.

Volont, Louis. 2019. "DIY Urbanism and the Lens of the Commons: Observations from Spain." *City & Community* 18 (1): 257–79.

Von Hassell, Malve. 2002. *The Struggle for Eden: Community Gardens In New York City.* Westport, CT: Praeger Publishers.

Warner, Coleman. 2001. "Freret's Century: Growth, Identity, and Loss in a New Orleans Neighborhood." *Louisiana History: The Journal of the Louisiana Historical Association* 42 (3): 323–58.

Weems, Carl F., Leslie K. Taylor, Melinda F. Cannon, Reshelle C. Marino, Dawn M. Romano, Brandon G. Scott, Andre M. Perry, and Vera Triplett. 2010. "Post Traumatic Stress, Context, and the Lingering Effects of the Hurricane Katrina Disaster among Ethnic Minority Youth." *Journal of Abnormal Child Psychology* 38: 49–56.

Weil, Frederick D. 2008. "NOLA-YURP Survey—July/August 2008 Preliminary Findings." http://www.lsu.edu/fweil/NOLA-YURP080823.pdf.

White, Monica M. 2011. "Sisters of the Soil: Urban Gardening as Resistance in Detroit." *Race/Ethnicity: Multidisciplinary Global Contexts* 5 (1): 13–28.

———. 2010. "Shouldering Responsibility for the Delivery of Human Rights: A Case Study of the D-Town Farmers of Detroit." *Race/Ethnicity: Multidisciplinary Global Perspectives* 3 (2): 189–211.

Wilkerson, Isabel. 2010. *The Warmth of Other Suns: The Epic Story of America's Great Migration.* New York: Random House.

Willson, Ian. 2018. "What Happened to the Hollygrove Market? An Interview with Nicola Krebill." Antigravity. July 9, 2018. https://antigravitymagazine.com/feature/what-happened-to-the-hollygrove-market-an-interview-with-nicola-krebill/.

Wittman, Hannah, Jessica Dennis, and Heather Pritchard. 2017. "Beyond the Market? New Agrarianism and Cooperative Farmland Access in North America." *Journal of Rural Studies* 53 (July): 303–16. https://doi.org/10.1016/j.jrurstud.2017.03.007.

Zukin, Sharon. 1982. *Loft Living: Culture and Capital in Urban Change.* Baltimore: Johns Hopkins University Press.

———. 2016. "Gentrification in Three Paradoxes." *City & Community* 15 (3): 202–7.

INDEX

Page numbers in *italics* indicate Figures, Tables, or Photos

expressed by, 74–75; as creative class, 223; cultivation sites dominated by, 11; for-profit projects led by, 84–85; as growers, 86; positionality reflected on by, 224; social entrepreneurial aspirations of, 82; in urban cultivation scene, 220, 223

trauma, secondary. *See* secondary trauma

Tremé neighborhood, 83, *193*

Tremé (television series), 42

Trouser House, 108–9, 141

truck farms, 27, 29, 269n9

Tulane University, 110, 255, 260, 262, 268n43; City Center of, 47, 270n56; Louisiana Research Collection at, 259

unhoused population, 172–73

United States (US), 8–9, 139, 153, 226, 235, 269n15; agricultural labor devalued in, 133; Census Bureau, 35–37, 39, 41; Department of Agriculture, 5, 102, 238, 241–42, 276n13; farm to table system in, 142–43; public trust declining in, 206; urban agriculture in, 238–39; wealth disparity in, 224. *See also* New Orleans

University of New Orleans, 47

unsettled times, 58, 87

Upper Ninth Ward: blight citation practices in, 180; St. Claud neighborhood in, 236; urban agriculture opposed in, 99

urban agriculture, 4, 156–57, 161, 178–79, 256; food insecurity not addressed with, 241; formalization of, 5, 238–39; metabolic rift and, 122; municipal policies incorporating, 238–39; popularity rise since the 2000s of, 6–8; social justice activism and, 6; Upper Ninth Ward opposing, 99; urban cultivation distinguished from,

3–4; urban planning incorporating, 238–39; US historical trends in, 5–6. *See also* gardening programs; policies

urban agriculture imaginary, 85

urban cultivation, 3–4, 178–79; in Black communities, 11, 219, 227, 269n33; children and, 99–100, 153–57, 192, 235; collective hopelessness and, 213; community rebuilding aspirations and, 43–44; co-optation and, 52; as cultural repertoire, 20, 87; cyclical popularity of, 20; definition of, 3–4; development threatening, 215; diminishing, 29–30; economic sustainability lacked by, 182; funding for, 241–42; gentrification and, 214–22; grassroots movements associated with, 227; growers leaving, 189–90; history of, 259–60; history of Black, 72; hope associated with, 14–15; interest in, 33–34; life stages and, 188; long-term residents resisting, 99; lots and types of sites for, 35; memory and history of, 25–31; next for, 238–44; nonprofit organizations supporting, 110; permaculture and, 60–61; in post-Katrina New Orleans, 9–11, 16–17, 34–40, 36, 37, 38, 39, 203; as prefigurative urbanism, 11–17, 64, 126, 133, 229, 233; as primary occupation, 162, 164–65; resources required to establish, 241; social issues addressed with, 232; social movements and, 263–64; universities supporting, 110; in Vietnamese immigrant enclave, 228. *See also* aspirations, growers; chickens; flower growing; goats; market activities; raised beds; urban cultivation scene; urban cultivation sites; urban farms

urban cultivation expansion aspirations, 64–69, 65, 135–36, 163, 237

ABOUT THE AUTHOR

YUKI KATO is Associate Professor of Sociology at Georgetown University. She is co-editor of *A Recipe for Gentrification: Food, Power, and Resistance in the City* (New York University Press).